JOHN DEERE

Publications International, Ltd.

Louis Weber, CEO
Publications International, Ltd.
8140 Lehigh Avenue
Morton Grove, IL 60053

ISBN: 978-1-68022-884-7

Manufactured in China.

8 7 6 5 4 3 2 1

Credits:

Contributing Writers: Marci McGrath, Doug Mitchel, Robert N. Pripps, & Chris Smith

Photography:

Back- and front-cover photos: www.deere.com

The editors would like to thank the following people and organizations for supplying the photography that made this book possible. They are listed below.

AGCO Corporation; Alamy Images: Malcolm Case-Green, Nigel Cattlin, Clynt Gamham Agriculture, Robert Convery, David R. Frazier/Photolibrary, Inc, Grant Heilman Photography, Doug Houghton, Kim Karpeles, Philip Lewis, RIA Novosti, Jim Ringland, George Robinson, Tim Scrivener, Doug Steley, Stephen Saks Photography, T.M.O. Pictures, Peter Titmuss, Jim West; AP Images; Art Resource: New York Public Library, SEF; Steve Ballard; Joerg Boethling: Agenda, Peter Arnold Inc.; David Bordner Collection; Corbis: Bettmann, Rick Dalton/AgStock Images, Gehl Company, Lake County Museum, Rick Miller/AgStock Images, PEMCO - Webster & Stevens Collection, Museum of History and Industry,Seattle, Louie Psihoyos/Science Faction, Ed Young/AgStock Images; Curt Teich Postcard Archive; Defense Industry Daily; Fernando del Real; David Drew; Robert Elzey; FarmPhoto. com; Planeta Gadget; Getty Images: Time Life Pictures; Greg Goebel; The Granger Collection; HA.com Photography; Hartland Equipment Corporation; High Contrast; Mark Igleski; Institute for Regional Studies and NDSU University Archives; The Image Works; IMRE Communications; John Deere Web site; KS Construction; Larson Farm & Lawn, Inc.; Library of Congress; M. Hanna Construction Co., Inc.; MachineFinder.com; Vince Manocchi; Glen A. Martin/Martin's BikeShop, Inc.; McLean County Historical Society; Doug Mitchel; Minnesota Historical Society: Duane Lundquist, Gordon Ray; Missouri Valley Special Collections, Kansas City Public Library; Andrew Morland; National Archives, and Records Administration; Nebraska Tractor Test Laboratory, Records, Archives & Special Collections, University of Nebraska-Lincoln Libraries; Photofest: Buena Vista; Picasa; PIL Collection; Platte Valley Equipment; Robert N. Pripps Collection; Andrew Raimist; Balaji Rengarajan; Rock Island County Historical Society; Orion Samuelson; SuperStock; Thomasnet News; Tractor Pool; Van's Implement, LTD; Wisconsin Historical Society; & Yale University Manuscript and Archives.

Owners:

Special thanks to the owners of the tractors featured in this book for their cooperation.

Howard "Shorty" Bonner; H.G. Bouris; Hershel "Junior" Conway; Kenneth Dutenhoeffer; Pete Dykestra; Michael Fondren; Ed Frichtl; Tom Garrison, Randy, Tammy, and Will Germany; Nick Guriel; Bruce Johnson; Larry Kindelsperger; Sims McNight; W. C. "Bill" Milligan; Jon H. Peterson; Paul Sawyer; Duane Schlomann; Romaine and Kathy Schweer; Randy and Sharon Sterwald; Billy Surrat; Danielle Talbott; Lance Talbott; Marilynn Talbott; Ray Volk; James Welsh; Stanley White; Ed Winkleman; & Gary Winkleman.

TABLE OF CONTENTS

INTRODUCTION

What began in 1837 with a plow made from a saw blade has since grown into one of the world's foremost manufacturers of farm equipment. "John Deere Green" has become almost a trademark in itself, and the products that wear it have become revered the world over.

For the first 80 years of its existence, John Deere built and sold only implements, leaving the fledgling tractor market to others. But as tractors became more accepted, it was apparent that farmers were looking for equipment compatibility and "one-stop shopping." Although John Deere had already developed its own tractor designs and invested in prototypes, it was decided that buying an already-proven machine and selling it through John Deere dealers would give them an instant product with a strong reputation. To that end, the company purchased the makers of the Waterloo Boy tractor in 1918, and began advertising and selling the large machine under the John Deere banner.

From then on, Deere has released a wide array of tractors and agricultural equipment that meets the needs of any sized farm and has become synonymous with American agriculture and the nation's heartland. John Deere has known no boundaries and continues to offer new and innovative products that solve the toughest agrarian problems.

Although this book focuses on the glorious tractors that brought mechanization to the nation's farms, let us not forget the hardy souls for whom they were built. A good tractor guided by the able hands of a hard-working farmer is a century-old partnership that has helped feed our nation—and the world.

THE DAIN MODEL

Joseph Dain was on the board at John Deere in 1914 when a majority of directors thought the company should explore the pros and cons of getting into the tractor business. Dain was asked to look into the concept further and report back to the board with his findings. Dain returned in 1915 with a prototype that he thought could be built and sold at a profitable price.

Although the board was lukewarm to the idea, Dain was given approval to continue testing. It appeared that his all-wheel-drive tractor could be sold for $1,200, which was an attractive retail price considering the advantages the machine could bring to farmers. After earning board approval to build 100 units of his new tractor, Joseph Dain suddenly died. Before that tragedy, Deere spent $250,000 to build just 10 units. Shortly after, the Dain project was removed from further corporate discussion.

1. The first Deere tractor to go into production was developed by board member and engineer Joseph Dain, Sr. The three-wheeled Dain machine improved on the short-lived Melvin design, a prototype tricycle tractor designed by staff engineer C.H. Melvin in 1912.

MODEL D

Although small quantities of largely experimental tractors were built earlier by John Deere, the company's first commercially successful tractor was introduced in 1923 as the Model D. And successful it was: Deere sold 23,000 Model Ds in its first five years on the market, and it continued with only minor changes for 15 years before it was "styled"—in which form it continued for another 15 years.

The most noteworthy feature of the Model D was its engine—not so much for its technical sophistication or power output as for its historical significance. Its two huge cylinders were laid flat in the frame facing forward, and during two revolutions of the crankshaft (720 degrees), the first cylinder fired at 0 degrees, the second at 180 degrees. Then the crankshaft would rotate $1^{1/2}$ turns (540 degrees) until the first cylinder fired again. Since the "space" between the two firing events was unequal, the engine produced an odd cadence as it ran. Not only would this basic engine configuration power most John Deere tractors for the next 35 years, its rhythmic beat prompted admirers to nickname the tractors "Poppin' Johnnies" or "Johnny Poppers," names the two-cylinder tractors are still lovingly known by today.

Early Model Ds can be identified by their six-spoke flywheels, which were replaced by solid flywheels at the end of 1925. The engine initially displaced a whopping 465 cubic inches, good for more than 22 horsepower at the drawbar (indicating how much power was available to pull an implement such as a plow) and 30 horsepower at the pulley (which was used to power machinery).

That was enough for the 4,000-lb tractor to pull three plows in most soil conditions. The engine was enlarged to 501 cubic inches for 1928, by which time it produced 28 horsepower at the drawbar and 36 at the pulley. By the end of the unstyled era, it was up to nearly 31 at the drawbar and more than 41 at the pulley.

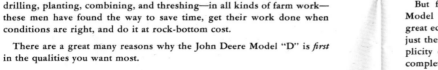

THE *Leader*

in all the QUALITIES
that mean GREATER Farm Profit

Power to pull a 4-bottom plow in most soils and three bottoms under practically all conditions—power to pull a big combine —power to operate a 28-inch thresher— simple, economical, dependable power for every farm job—that's the famous John Deere Model "D" Tractor.

From the Atlantic seaboard to the west coast, from the Peace River country in Canada to the Gulf of Mexico, and in all parts of the world where standard tread tractors are used, it is the universal opinion of the many thousands of owners that the John Deere Model "D" Tractor is the leader in all the qualities that mean greater farm profit. Plowing, disking, drilling, planting, combining, and threshing—in all kinds of farm work— these men have found the way to save time, get their work done when conditions are right, and do it at rock-bottom cost.

There are a great many reasons why the John Deere Model "D" is *first* in the qualities you want most.

FIRST IN SIMPLICITY

Only John Deere tractors give you the simplicity of exclusive two-cylinder engine design . . . fewer parts . . . a straight-line transmission without power-consuming bevel gears . . . a belt pulley right on the crankshaft, the simplest construction possible.

FIRST IN DEPENDABILITY

Only John Deere two-cylinder tractors give you the dependability of fewer and sturdier parts . . . greater ability to stand up under heavy loads . . . proper distribution of weight for better, more positive traction.

FIRST IN ECONOMY

Only John Deere two-cylinder tractors are backed by a long-time record of success, efficiency and safety in burning the low-cost fuels such as distillate, furnace oil, fuel oil, stove tops, Turner Valley naphtha, and some grades of Diesel oil . . . fuels that cost far less and are approximately 10% more powerful . . . fuels that the John Deere converts into steady, dependable power on drawbar and belt. Burning the low-cost, money-saving fuels in a John Deere tractor is *no experiment.*

But fuel economy is not all. John Deere Model "D" owners also benefit from another great economy . . . the ability to inspect and adjust their tractors right on the farm. The simplicity of two-cylinder design makes possible complete accessibility, one of the big reasons why a recent survey shows that 82% of John Deere owners do fully 75% of their own service work.

EXTRA-LONG LIFE

To the other great features listed, add the extra years of service that the John Deere Model "D" gives, to better understand the big swing to John Deere two-cylinder power. Fewer, heavier, longer-lived parts . . . the use of high-quality materials, careful workmanship, and rigid inspection . . . make records of 8, 10, and even 12 or more years of service not at all unusual.

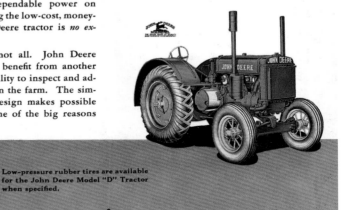

Low-pressure rubber tires are available for the John Deere Model "D" Tractor when specified.

- 2 -

- 3 -

1. Deere introduced the Model D in 1923, little imagining that the tractor, in unstyled and styled iterations, would enjoy continuous production for 30 years. The spoked flywheel and fabricated front axle identify this 1923 D as one of the first 50 made.

1. The Model D was created to supplant the Waterloo Boy Model N, a kerosene-powered tractor acquired by Deere in its 1918 purchase of Waterloo, Iowa, company Waterloo Gasoline Engine Co., and the two models were sold simultaneously for one year. The D was powered by a fuel-sipping two-cylinder motor. The tractor's stout configuration made it a jack of all trades that could be used for a variety of farming chores. The D would go on to become the longest running Deere of all time.

2. The perch for the operator of a prewar Model D hung far off the back section of the frame, but full-coverage fenders kept most of the flying debris a safe distance away. This early example wears the all-steel wheels that were the industry norm before widespread use of rubber tires. Incremental changes were made to the D throughout the model's lifespan, with each alteration adding more versatility to the already-popular model.

From its introduction in 1923 as the first tractor to wear the John Deere name to its final edition 30 years later, the Model D changed surprisingly little. One area in which it did change was in appearance. Like many other models in the line, the Model D received stylish sheet metal for the 1939 model year. Unlike the others, however, its new grille featured vertical slots rather than horizontal "gills," which instantly set it apart.

What didn't change—even with the new look—was the D's chunky, massive character. It was lower and much shorter than the row-crop Model A, yet also significantly heavier. With its huge 501-cubic-inch two-cylinder engine, the Model D was rated at nearly 31 horsepower at the drawbar and 38 at the pulley running on kerosene, and tipped the scales at a hefty 5,300 lb. None of these figures seem very impressive by today's standards, but by an early postwar yardstick, they were brutish.

A diesel-powered version of the Model D was introduced for 1949 as the Model R. It was even heavier and put out more horsepower, marking the first time the Model D wasn't the king of the John Deere hill. The assembly line churned out its last Model D in March 1953, but that wasn't the end of production. Several more were hand-built at a location away from the factory using spare parts. Like other tractors in the John Deere line, the Model D's redesigned successor received a numeric designation, arriving in late 1953 as the Model 80.

JOHN DEERE MODEL "D" TRACTOR

The Economical Tractor
for the
HEAVIER FARM JOBS

1. The Soviet hunger for tractors began with collectivization of farms in the 1920s, and led to Deere making a deal with the Soviets, selling them between 1,500 and 2,200 Model D tractors. By 1931, the Soviets had bought $9.8 million worth of Deere products. Soviet farmers and bureaucrats were satisfied that the D worked reliably and efficiently, and with minimum maintenance.

2. A padded seat that allowed for lower-back support made life with this 1948 D considerably easier than with earlier models.

MODEL GP

John Deere's first tractor, the Model D, was considered a fairly heavy-duty tractor for its day. Its 4,000-lb heft, 30 horsepower at the pulley, and low, wide stance made it perfect for plowing large fields and operating threshing machines off its pulley. But not every farmer needed such a large tractor, and, in fact, those who had them often found it necessary to use horses for more delicate work such as planting and cultivating. As a result, John Deere brought out the smaller Model C in 1927, which was renamed the Model GP (General Purpose) the following year. The GP weighed in at about 3,600 lb, and its engine produced about 20 horsepower at the pulley.

In addition to the pulley, it offered a power take-off (PTO) for running implements. Due to its lighter weight and lower cost, the GP sold in far greater numbers than the D and really helped put John Deere on the map. It was soon available with a narrow or wide front track, and it became the first John Deere to offer Power Lift, a power equipment lift. Initial models were fitted with traditional steel wheels; the factory didn't make pneumatic (rubber) tires available until the early 1930s.

1. The GP was named for its general-purpose design that aimed to satisfy all types of farming needs. A two-bottom plow could be pulled behind, giving the GP the ability to plant and cultivate three rows of crops in a single pass. This 1930 model displays the large radiator and stout design that were well suited to rugged duty. The 1930 GP was produced in standard and wide-tread variations; all GPs were fitted with three forward gears and one reverse.

1. Despite the success of its Model D, by the mid-1920s Deere had fierce competition for the row-farming market from International Harvester's hit model, the Farmall. The experimental 1927 Deere Model C, designed to pull a multiple-row planter attachment and perform other tasks, was intended to meet the all-purpose needs of small row farmers. Because of technical challenges and disagreements at Deere over small-tractor technology, only a few Model Cs were produced. However, the effort led to the emergence of the more successful GP model, which Deere produced until the mid-1930s.

2. In the 1930s, Deere's Model GP, or "General Purpose," tractor added improved horsepower to the versatility of earlier Deere all-purpose models. But by this time the devastating effects of the Great Depression were taking a toll on sales of all tractors. To compete, especially in the growing California market, Deere combined its sales effort in the mid-1930s with Caterpillar, then the leader in tracked-tread tractors.

3. Deere's GP was heavier and more sophisticated than the earlier Model D, but needed repairs in the field more frequently than should have been necessary. Between that and the onset of the Great Depression, GP sales never measured up to Deere's projections. The view from the GP's catbird seat was utilitarian, and because comfort wasn't yet a priority, the drilled steel saddle was austere. The access panel was clearly marked to show manufacturer and place of assembly. The GP seen on this page is a late-1929 model, as evidenced by the vertical air intake that protrudes from the cowl.

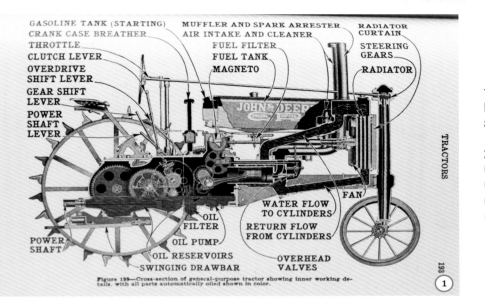

GASOLINE TANK (STARTING) MUFFLER AND SPARK ARRESTER RADIATOR CURTAIN
CRANK CASE BREATHER AIR INTAKE AND CLEANER
THROTTLE FUEL FILTER STEERING GEARS
CLUTCH LEVER FUEL TANK
OVERDRIVE SHIFT LEVER MAGNETO RADIATOR
GEAR SHIFT LEVER
POWER SHAFT LEVER

WATER FLOW TO CYLINDERS FAN
RETURN FLOW FROM CYLINDERS
OIL FILTER
POWER SHAFT OIL PUMP
OIL RESERVOIRS OVERHEAD VALVES
SWINGING DRAWBAR

Figure 199—Cross-section of general-purpose tractor showing inner working details, with all parts automatically oiled shown in color.

1933

1

1. An early Deere General Purpose (GP) tractor was given the cutaway treatment in this Deere handbook for farmers published in around 1932. Parts shown in color were oiled automatically. Deere promised that the GP "replaces horses rapidly and safely."

2. Beginning in the late 1920s, the Lindeman firm of Washington state converted Deere's GP to a crawler for use on the steep slopes of some orchards. Only 24 examples of the GPO (*O* for Orchard) were constructed, and soon after, the Model B was selected for conversion to BO models. Deere would later buy Lindeman and move that company's operations to Iowa for convenience and efficiency.

2

Model GP 19

MODEL A

As tractors became more accepted, they also became more diverse. Some were developed to be better-suited to specific tasks, and the Model A that arrived in 1934 is a good example.

The Model A is considered a "row-crop" tractor, as its front wheels are close together and angled slightly (to go between two rows) and its rear wheels can slide in and out on their axles so they can be adjusted to straddle rows of various widths. In size, weight, and power, the A was larger than the GP, which evolved into the Model B. The A weighed about 3,800 lb and was rated at 24 horsepower at the pulley. Whereas most GPs had a steering column that angled down toward the engine, the A's column ran almost horizontally above the hood to a vertical steering post. The Model B, which arrived shortly after the A, had the same arrangement. They differ, however, in that on the A, the cover at the top of the post is smooth in front with four bolts on the back; on the B, the cover has two bolts on the front. The Model A has the distinction of being—by a narrow margin—the most popular two-cylinder tractor ever sold by John Deere.

One of the major innovations for tractors arrived in the early 1930s in the form of pneumatic rubber tires. Tests showed these "air" tires transferred more power to the ground than steel wheels. But even more important to farmers was the fact they provided a far smoother ride, and also allowed the tractors to be driven on streets, which meant they could be used to pull wagons into town.

1. By 1933 overall industrial output across the USA had been cut in half contrasted to pre-1929 levels. General unemployment flirted with the 25 percent mark. Prices, especially farm prices, plummeted. It was in this perilous context that Deere president Charles Wiman made a fateful decision. He instructed Theo Brown, the company's research manager, to aggressively pursue the development of two new general purpose tractors. For Wiman personally, the program allowed him to express his appreciation for machinery and his training as an engineer. But if he was wrong about the eventual success of these new tractors, the company could go bankrupt. Wiman's plan called for two machines: a two-plow, the Model A; and a one-plow version called the Model B. These were designed as replacements for the disappointing GP and to counter the successful Farmall and Oliver "row-crop" tractors. This unstyled 1935 Model A still sports "General Purpose" markings.

1. The Model A came out first, in April 1934, followed by the B almost a year later.

Like the smaller Model B, John Deere's Model A became "styled" for 1938 by New York industrial designer Henry Dreyfuss, who was hired by Deere to breathe a breath of modernity into Deere's rugged tractors. Also like the B, the process brought sleeker lines headed by a sheet-metal grille, but in this case, the grille had eight slots instead of seven. (The larger Model G also had eight, but the two can be told apart because the G had its air-cleaner and exhaust stacks side-by-side, whereas the A had them one behind the other, like the B.)

Despite the obvious appearance improvements, ads of the day continued to stress the low cost of John Deere ownership. "Upkeep is mighty low," one states, and "The John Deere is the simplest tractor built … But where you will notice the big saving is in fuel costs." Both benefits were credited to the tractor's two-cylinder engine, a design Deere had been building for 20 years.

By this time, the Model A was rated at 26 horsepower at the drawbar, nearly 30 at the pulley. The two figures were much closer than they had been 15 years earlier (see the GP) due to advances in driveline technology that reduced power loss. Also, "pulley horsepower" was becoming a dated term; most tractors could be ordered with a power take-off (PTO), and that was rapidly becoming the standard way to drive equipment—and measure horsepower.

1. 1945 Model A, with drilled, saddle-style seat.

After getting styled for 1938, the Model A received a Powr-Trol hydraulic system in late 1945. Powr-Trol powered a hydraulic lift cylinder and allowed precise positioning of a towed implement. The next big step in the evolution of the Model A occurred in 1947, and this general design would carry through to the end of production. Most noticeable was the new pressed-steel frame, which made these models instantly identifiable due to a "kick up" just behind the grille that covered the side of the engine. A padded seat was also available, and the battery was placed beneath it. A slightly larger 332-cubic-inch engine with standard electric starting was fitted to these "late" Model As, and horsepower was measured at 38 at the pulley/PTO—quite a step up from the 16 horsepower produced by the earliest versions. But weight had also increased substantially since the model's introduction in 1934, nearly doubling to about 6,000 lb.

There's a *Lot of Tomorrow*
in Today's **JOHN DEERE TRACTORS**

1. With production beginning in 1934, the Model A had already enjoyed a long life when this 1948 example showed up at dealerships. Over the years, revisions and upgrades had happened on a regular basis, with the most obvious being the Dreyfuss-styled redesign of 1938. Operator comfort became more important after the war, so the earlier pan-shaped steel seat was replaced by a padded unit that spared the operator's rump and lower back.

TODAY—in a great line of tractors—John Deere is setting the pattern for tomorrow's farm power.

There's a new *get-up-and-go* to provide the extra power you've always wanted to match your heaviest jobs, to speed up every job and save time. There's effortless, finger-tip control to raise and lower drawn or mounted implements, to change their operating position instantly for best work. There's a new freedom in steering, a new smoothness over rough ground that eliminates wheel-tug and greatly reduces fatigue. There's a new ease of attaching and detaching cultivators and many other mounted implements that cuts change-over time.

These are just a few of tomorrow's advantages you can enjoy *today* in a John Deere—advantages made possible by advanced engineering. Add to them the outstanding economy, the rugged dependability of exclusive John Deere *two-cylinder* design with its fewer, heavier parts ... the quality construction that is typical of John Deere ... and you'll understand why John Deere Tractors are *first* in *modern* design and *proved* performance.

JOHN **DEERE** *Moline, Illinois*

CYCLONIC FUEL-INTAKE ENGINES

HYDRAULIC POWR-TROL

TOUCH-O-MATIC CONTROL

QUIK-TATCH EQUIPMENT

ROLL-O-MATIC KNEE-ACTION FRONT WHEELS

FALGREN TRACTOR MARCH 1948

49

LEONARD FAHLGREN

1.The tricycle configuration was the conventional arrangement for general purpose machines at the time, with the two front wheels sitting close together. By 1937, variations with only one front wheel, with adjustable wide-fronts, and extra high-clearance (Hi-Crop) versions of all front-end types were made available.

2. On a snowy patch in the late winter of 1948 a Deere Model A tangled with a 1940 Chevrolet. Apparently, nobody was hurt, but the Deere's starboard exhaust stack—as well as the Chevy's front end—took a beating.

During its 20-year life span, numerous versions of John Deere's venerable Model A were offered, which helped cement its position as the best-selling tractor in the company's history.

One of the more unusual configurations was the Hi-Crop version. The example pictured below was originally shipped to California for use in gladiola fields.

Raising the front end off the ground was fairly easy, but raising the rear required special gearboxes that were fitted between the axle and the wheel hub. Easing entry to the elevated helm were step-plate "stairs" mounted to the left-rear axle housing.

When the Model A was finally retired in 1953, it was succeeded in the line by the Model 60, which didn't enjoy anywhere near as long a life.

1. This Model A Hi-Crop is from 1952, the final year of Model A production.

2. A styled Model A heads a Bob Wills parade in Turkey, Texas, around 2002.

MODEL AO

Shortly after the 1934 introduction of the Model A row-crop tractor, John Deere brought out a version specially suited for use in orchards. It was—quite appropriately—designated the Model AO. Visually distinctive features of the AO were its standard-tread (wide) front axle that permitted a low ground clearance, stubby air-cleaner and exhaust stacks, and wide, sweeping rear fenders; fenders that fully enclosed the rear wheels were also available (see ad). All these modifications were made to avoid "catching" low-hanging branches.

Also featured on the AO were individual rear brakes. These allowed the operator to brake just the inside wheel during a turn, which made for a tighter turning radius—useful when trying to maneuver around closely spaced trees. The AO was one of the few John Deere tractors that didn't get styled in the late 1930s or early 1940s. Instead, it soldiered on in unstyled form, as shown on this featured 1944 version.

1. The Deere AO, which went on sale in 1935, was configured for use in orchards. It was based on the successful Model A, but with enclosed fenders that protected the low branches of the trees being cultivated. By 1944 the AO carried a 321-cubic-inch, two-cylinder engine, and was capable of 12.3 mph in sixth gear. Reverse allowed a top speed of 4 mph. All AO engines ran on gasoline. The AO's vertically sited steering wheel did what it needed to do, but failed to consider the comfort of the driver.

MODEL B

Introduced for the 1935 model year, the Model B effectively replaced the Model GP, which had been around since 1928. The Model B weighed about 2,800 lb and was rated at 18 horsepower at the pulley. Like its bigger Model A brother, the B was offered as a row-crop tractor, with its front wheels close together and its rear wheels able to be moved in and out on the axle to vary the distance between them. The design allowed the front wheels to go between narrow rows, the rear wheels to straddle rows of different widths. The B was also offered with wide-spaced front wheels, which, like the rears, could be adjusted in and out to vary the track width.

On the very first Model Bs, the steering post was attached to the frame with four bolts; soon after, eight bolts were used. As a result, four-bolt models are very rare and quite valuable. During its life span, which extended through 1952, the Model B sold only slightly fewer copies than John Deere's best-selling two-cylinder tractor, the Model A.

1. More than 300,000 Model Bs would be built in-between the years 1935 and 1952.

2. The Model B was originally available with pneumatic tires. Like the A, it had a four-speed transmission, a power take-off (PTO) setup, and belt pulley. Its engine, a scaled version of the A's, had enough power for two 16-inch plows while the A was capable of pulling four 16s.

The New Model B

MEET THE SMALLER BROTHER OF THE FAMOUS JOHN DEERE MODEL A

WHEN the John Deere General Purpose Model A, the tractor with the adjustable tread, was introduced last spring, it met with immediate farmer acceptance—it proved to be the greatest forward step in row-crop tractor design since the inception of the tractor.

And rightfully so, for here was a farm tractor developed on row-crop farms and built to meet the most exacting needs of row-crop farmers. The Model A proved a revelation in ease of handling, in its adaptability to varied crop requirements, and above all, in low-cost operation.

Farmers Asked for a Smaller Model A

But farms vary greatly in size—many farmers do not need the big capacity of the Model A in plant-ing and cultivating, and other farm work. As a result, among the smaller farmers this remark was often heard—"Give me a smaller tractor just like the Model A that will enable me to do away with horses and I'll start farming with a John Deere Tractor and equipment tomorrow."

And larger farmers said: "In addition to our present power for the heavier work, we can use a duplicate of the Model A, only in smaller size, as supplementary power."

To Meet That Demand the Model B Is Ready for You

A smaller tractor—about two-thirds the size of the Model A in power, weight, and dimensions—a tractor that pulls one 16-inch bottom, a two-row cultivator, and other machines requiring proportionate power—

—but a tractor with all of the nine outstanding features of the Model A.

Get Acquainted with these New General Purpose Tractors

Call on your John Deere dealer and investigate these unusual tractor values. Let him show you the nine superior features found in both the Models A and B. At the first opportunity arrange for a demonstration—get on the seat and drive these tractors yourself. Only then can you fully appreciate their smooth, responsive power, how easy they handle, the perfect view you have of the work, and their adaptability to your many farm jobs. Fill out the coupon below or write to John Deere, Moline, Ill.—state the tractor in which you are interested and ask for booklet B-38.

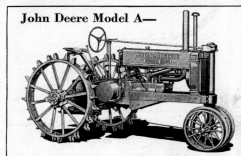

John Deere Model A—

—The Larger General Purpose for use on average size and larger row-crop farms—pulls two 14-inch bottoms, a two- or four-row cultivator, an 8-foot double-action disk harrow, a 22-inch separator and does many other farm jobs within its greater power range. It is also the preferable size for many smaller farms having hilly or difficult soil conditions.

FILL OUT AND MAIL THIS COUPON

JOHN DEERE Moline, Ill.	B-38
Please send me special literature on the John Deere tractors I have checked.	Name....................
☐ Model A	Town....................
☐ Model B	State................ R. F. D........

SUCCESSFUL FARMING, *February, 1935*

1. Deere introduced the Model B in 1934 as a smaller alternative to the A. This is an unstyled 1935 B.

2. A styled Model B, with jury-rigged spotlight, works the fields after dark.

In late 1937, the Model B was given a longer frame that allowed it to share more parts with the larger Model A, thus cutting production costs. In collector circles, these later Bs are known as "long frame" versions.

Model Bs were offered with a variety of front-end designs. "Wide-tread" models (denoted with a "W" in the model name, such as "BW") had the front wheels far apart, while "tricycle" versions had them close together (also called "Narrow tread," which prompted an "N" in the model designation). A variation of the tricycle had a single front wheel, as shown here. Some models of both Wide and Narrow configuration were mounted higher off the ground for more crop clearance, and these added an "H" to their model name.

1. The Model Bs were immensely profitable for both Deere and the farmers who bought them. The years brought a continual process of upgrade and improvement. One great line of demarcation occurred in 1938, however, when styling came to the John Deere tractor.

2. One of the selling points of the Model B was its fuel economy, much of which was attributable to its two-cylinder design. (Many competitors used thirstier four-cylinder engines.) Another was that some versions could be run on low-grade fuels such as distillate and furnace oil, which cost less than gasoline. Since fuel made up a large part of a farmer's budget, these were important considerations—something ads of the day clearly pointed out. Also mentioned were individual brakes on the rear wheels that allowed for tighter turning and a hydraulic power lift for implements. In 1931, it was estimated that only one farmer in six owned a tractor, despite studies that showed tractors cost less than horses to maintain. But as the Depression eased in the mid-1930s and tractor costs came down, more and more farmers made the switch. With its low price, low operating costs, convenient features, and strong reputation, the Model B was a popular first tractor for farmers who were "moving up" from horses.

After the advent of pneumatic tires in the early 1930s, the next great leap in tractor development came with the "styled" tractors that arrived later that same decade. In the case of John Deere, most of the changes can be attributed to industrial designer Henry Dreyfuss. Dreyfuss did not work for Deere, but rather had his own design firm and operated out of New York.

The first styled John Deere tractors appeared for the 1938 model year. Sheet metal surrounded the radiator, flowed into the side panels, and enclosed the top-mounted steering column, providing a sleek, more modern look. Styled Model Bs were given seven radiator slots, and their exhaust stack sprouted from the hood just ahead of the air-cleaner stack.

However, the beauty of these tractors was more than skin deep. Ergonomics (the relation between seat, steering wheel, pedals, and controls) were also improved, which—along with the "cushy" pneumatic tires—greatly increased operator comfort and productivity.

Despite the advantages in both appearance and comfort, not all John Deere tractors got styled. An ad from the early 1940s shows that several models, including the big Model G row-crop tractor and the Orchard variants, escaped Dreyfuss's touch. Also shown is the depth and variation of the John Deere lineup at this point, with sizes and styles to meet almost any farming need.

1. In 1936, just two years after the debut of the Model B, world-class designer Henry Dreyfuss was hired to freshen the looks of Deere's workhorse series. The altered sheet metal made an immediate sales difference, as farmers were drawn to the sleeker style of new B models that came out in 1938. This 1941 B has the revised grille and hood that came from the Dreyfuss studio. The 1941 Bs also offered a six-speed gearbox for tractors equipped with rubber tires. Steel-wheel variants had to make do with the previous four-speed box.

2. Of all the variations seen for the Model B, this one, the BWH, is one of the rarest. Due to the specific nature of its usage as a Hi-Crop machine, only 50 examples were produced. The wide spacing of the front wheels and exceptional clearance made the BWH virtually purpose-built for corn, cotton, and sugarcane. This example is one of the first "styled" tractors to be built by Deere.

MODEL BO

As had its Model A counterpart, the Model B was offered in a specially designed version for use in orchards. Called the Model BO, it featured many of the alterations of its AO sibling, including a low ride height, ultra-short air cleaner and exhaust stacks, and encompassing rear fenders, all in the interest of avoiding damage to low-hanging branches. Also available were fully skirted rear fenders.

Like the AO, the BO had individual rear brakes that allowed the operator to brake just the inside wheel for tighter turns in the confined space of an orchard. Also like the AO, the BO didn't get the late-1930s styling treatment applied to most other models in the John Deere line, instead remaining in unstyled form through the end of its production run.

John Deere Model "AO"
Grove and Orchard Tractor

The John Deere Model "AO" Grove and Orchard Tractor has the pulling capacity of a 6-horse team, and the daily work output of 8 horses. At its highest point, the top of the cowl, this tractor stands only 53 inches high. Ideal for orchards, groves, vineyards, hop-yards.

John Deere Models "AO" and "BO" Grove and Orchard Tractors

ONE glance at a John Deere Model "AO" Grove and Orchard Tractor and you know that here is a tractor specially designed for orchard work . . . built low and completely streamlined to save trees and fruit.

One trip into the orchard and you are more convinced. Short turns around trees, four forward speeds, built-in power shaft, easy handling . . . these are all important features.

One day's work and you *know* that the John Deere is the tractor you want. John Deere economy wins again.

Both the fully streamlined Model "AO", and the slightly less streamlined Model "BO", are built low and are fully protected to avoid catching branches. Citrus fenders are special equipment. On rubber-tired models, solid cast wheels eliminate all rear-wheel spokes and, in most cases, make wheel weights unnecessary. Both tractors work in close to trees, even though branches are trimmed low.

Easy Handling in Tight Quarters

Due to automotive-type steering, and independently operated differential brakes, you can make extremely short turns around trees and at the ends of tree rows. There's plenty of room for the operator on the platform, and the seat is so located that the operator's head is only a little above the steering wheel.

Handle Many Jobs

Four speeds forward—2, 3, 4, and 6-1/4 miles per hour—adapt these tractors to a wide variety of uses including hauling. There is a reverse of 3 miles per hour. You'll find these tractors efficient power plants for both field and belt work. In plowing, you have a center hitch to both plow and tractor, with two wheels in the furrow.

John Deere Models "AO" and "BO" Tractors have all the mechanical advantages that make John Deere tractors so economical, so dependable, so adaptable, so outstanding.

John Deere Model "BO"
Grove and Orchard Tractor

Standing only 52 inches high, the John Deere Model "BO" Grove and Orchard Tractor handles the load of a 4-horse team, and provides the sustained work output of 6 horses. It is shown at right with steel wheels which are regular equipment, and with special citrus fenders which are an extra. Can also be furnished with low-pressure rubber tires and solid cast wheels as shown on the Model "AO" Tractor, above, at extra cost.

1. Operation of a typical tractor in sandy soil could be almost impossible, so Deere needed a machine suited to work farms with those conditions. Hence the tracked crawlers, like this 1947 Model BO, which was converted from wheels to treads by the Lindeman Power Equipment Company in Yakima, Washington. Conversion nearly doubled the cost of the tractor, but the tweaked Deere model was well received by those who needed a broader footprint. When Deere elected to phase out the BO model after World War II, Lindeman anticipated a devastating decline in revenue because of the complexities involved in conversion work performed on tractors other than the B. To preserve Lindeman's expertise, as well as ensure the continuation of reasonably priced conversions, Deere bought Lindeman in 1947 and moved the company to Dubuque, Iowa.

2. An interesting variation of the Model BO was a tracked crawler built by Lindeman, a company otherwise not associated with John Deere. On Lindeman versions, the steering wheel was replaced by a pair of control levers. These activated track clutches that would transfer power to one side or the other, causing the tractor to turn—a system similar to that used on tanks. John Deere purchased Lindeman in the late 1940s, and soon came out with its own crawler based on the upcoming Model 40.

MODEL H

In an effort to appeal to farmers for whom even the Model B was a financial stretch, John Deere brought out the Model H in 1939. It arrived in styled form and looked much like a scaled-down B. It had seven grille slots (like the B), but only the exhaust pipe poked up from the top of the hood; the air cleaner was housed behind a small, mesh screen on the left side-panel, just ahead of the John Deere logo. Weighing in at 2,063 lb, the Model H was about 700 lb lighter than a Model B. It also had a much smaller engine: 100 cubic inches vs. 149, which would prove to be the smallest horizontal twin John Deere ever built. It was rated at 12.5 horsepower at the drawbar, 15 at the pulley. John Deere called the Model H a "one-to-two-plow" tractor, which slotted it between the little one-plow Model L and the two-plow Model B. (The larger Model A was rated for two-to-three plows, the big Model G for three plows.)

In addition to its "first-time buyer" audience, the Model H was also marketed as a light-duty workhorse on farms that already had a larger tractor. Ads promoted both uses, noting that the Model H "handles all jobs on the small farm, replaces the last team [of horses] on the large one." Indeed, some farms that were "mechanized" at the time still used horse-drawn equipment for some tasks—but that was quickly changing.

1. The Model H was introduced in 1939, and was aimed squarely at farmers who worked spreads of fewer than 80 acres. The H was a capable tractor with plenty of flexibility. It was available in five configurations ranging in price from $595 to $650. Rated at a maximum belt horsepower of 14.2, the H was well suited for those smaller farms. Deere built the H between 1939 and 1947, with total production approaching 59,000 units.

MODELS L/LA

The smallest tractor John Deere offered during the "Two-Cylinder" era was introduced in 1937 as the Model 62, but was renamed the Model L later that same year. It was powered by a vertical two-cylinder engine built by Hercules that developed seven horsepower at the drawbar, 10 at the pulley, enough to warrant a one-plow rating. Weighing in at just 1,515 lb, it featured a foot-operated clutch (most Deere clutches were operated with a hand lever) and individual rear brakes that allowed a seven-foot turning radius, making it perfect for use in tight quarters. Styled sheet metal was applied to the Model L in 1938, and it got a Deere built engine with the same power rating for 1941. It was joined that same year by the slightly larger LA, which looked nearly identical but tipped the scales at 2,200 lb, and its larger engine produced 13 horsepower at the drawbar, 14 at the pulley. Production ceased on the L and LA midway through 1946, and no successor replaced them in the lineup.

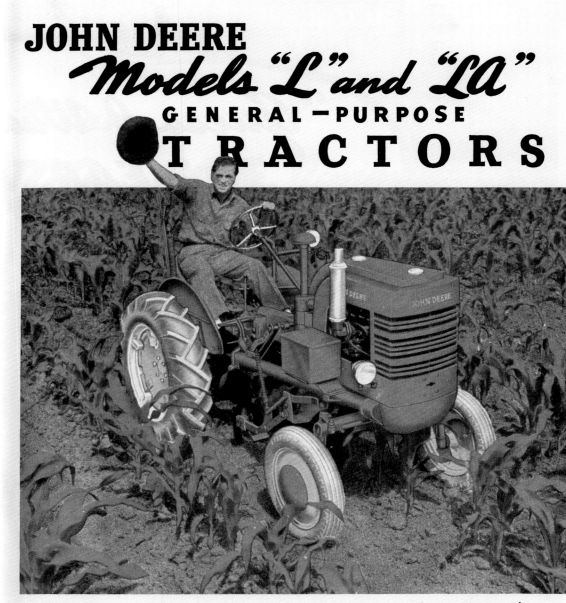

1. Domestic sales of tractors plummeted during the Depression, so Deere and other manufacturers felt pressed to come up with low-priced machines that would prove their worth in the fields. Deere's answer was the Model L, which appeared in 1937, simultaneously with the (1937-only) Model 62. Early L models were unstyled, with an exposed radiator. From 1938 until production ended in 1946, styled versions carried a shapely radiator enclosure and a curved cowl. The L was powered by a vertical, two-cylinder motor fed by gasoline only. A foot-operated clutch allowed the operator to choose among three forward gears. In all, 13,365 copies of the L were produced, selling for between $475 and $517.

1. In 1936, engineers at Deere's Moline Wagon Works had an idea for a tractor to replace a single horse for jobs on very small farms, or for a chore tractor on larger farms. Taking a clean-sheet-of-paper approach, the Moline team settled on a small, conventional running gear with a wide front end, a straight axle, and downward extending kingpins (for crop clearance). This arrangement would eventually replace the row-crop configuration and would be known as the "Utility" configuration. A tubular frame was devised and a two-cylinder Novo engine was mated to a Model A Ford transmission. The steering gear and steering wheel were also Model A Ford.

The use of Ford parts greatly speeded the development process of this "Model Y" prototype tractor. The Model Y was refined into the Model 62, which used a more powerful Hercules two-cylinder engine and a Deere-built transmission and steering mechanism. A few of these were sold to the public. In the 1937–38 model years, a further improvement to the 62 resulted in another designation; the Model L. In late 1938, the L received the Dreyfuss styling touch. Then, starting in 1940, the more powerful and heavier Model LA, with a Deere-built engine and its own serial number series, was introduced.

2. A styled 1941 Model L, with single-bottom plow.

1. Henry Dreyfuss designed the styled L models to look markedly different from other, larger Deere tractors. This one, with optional lamps, dates from about 1940.

2. The Model LA was introduced for 1941, and was built simultaneously with the L until both models departed. The saddle of this 1946 LA has been carefully fitted with padding.

3. This Model 62 was built during 1937. It was the second of a follow-on series of three small row-crop tractors that began with the Model Y in 1936, and culminated in 1937 with the Model L. All Model 62s ran with a 10.4-horse Hercules engine. This one sports full fenders.

4. During the mid-1930s, John Deere Wagon Works engineers worked to develop a new, low-cost tractor product. The result, a few dozen units of the promising Model Y, was renamed the Model 62 (shown here) in 1937. The rare departure from letter-names for tractor models was reversed a couple of years later when the design was dubbed the Model L. That tractor, which added styling improvements by Henry Dreyfuss, sparked a wartime surge in Deere sales.

MODEL R

When it was introduced for the 1949 model year, the diesel-powered Model R replaced the venerable Model D as top dog in the John Deere line. Its big 415-cubic-inch diesel engine continued the Deere tradition of horizontally mounted twin cylinders, and put out 45 horsepower vs. 38 for the kerosene-burning Model D.

Since diesel engines were notoriously hard to start—particularly in cold temperatures—the R was fitted with a small, two-cylinder, gasoline-powered "pup" motor that was mounted near the flywheel and used for starting. The pup motor had an electric starter, and once running, it could warm the diesel engine and then turn it over. Aside from its engine, the Model R was much like the Model D. Both came in only one configuration, with standard tread and a low ride height. The R was the first Deere model to offer an all-steel cab, along with "live" (meaning it was powered directly off the engine) PTO and hydraulics.

As it turned out, the Model R's reign was rather short. Along with most other "Letter Series" tractors, it lasted only through 1953, after which it was reborn as the Model 80—which likewise topped the line.

1. Deere's first diesel tractor was the Model R, introduced in 1949. It was based on developments that Deere had undertaken as early as 1940, with eight experimental MX models that were tested and then redesigned to address shortcomings. It's interesting to note that the final engine design for the Model R's two-cylinder diesel employed the same bore, stroke, and valve sizes as the six-cylinder Caterpillar D8. Some historians believe that informal cooperation between the two Illinois companies continued after Deere's 1945–47 move into industrial equipment.

2. The 1949 Model R was Deere's first foray into the world of diesel power. The pluses of diesel economy and torque far outweighed the negative issues of weight and hard starting in cold climates. A small two-cylinder, gas-powered "pony" engine was used to bring the 416-cubic-inch diesel motor to life. Although the R weighed in at nearly four tons, it could nevertheless reach 11.5 mph when rolling on rubber tires in fifth gear. Steel-wheel R models had a factory-blocked fifth gear, to limit top speed to 5.3 mph.

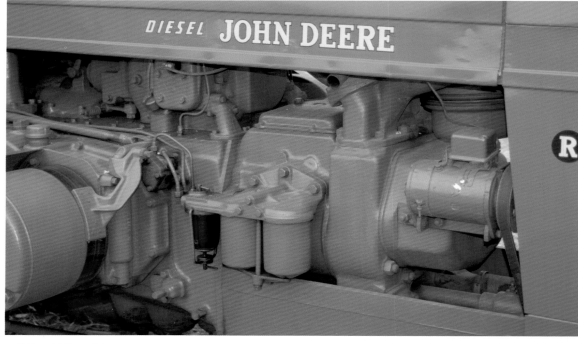

1. Deere first offered diesel power in 1949, on the Model R. Because diesels were famously reluctant to start in cold weather, Deere's unit was preheated by a gas-powered "pony motor" (which was itself electrically started). Dependable starts are central to this 1955–56 ad for Deere's Model 80 tractors.

MODEL M

Like its Model L and LA predecessors, the Model M of 1947 departed from usual John Deere practice in having its two-cylinder engine standing upright, mounted longitudinally in the frame; all other Deeres since the company's beginning had their engines laid flat with cylinders facing forward.

The M also used the engine as a structural member, making it a "unit design." Several new features marked the Model M. The padded seat included inflatable cushions and was adjustable fore and aft, while the steering wheel could be telescoped through a one-foot range, allowing the driver to either stand or sit.

Originally introduced with a standard front end, the Model MT (for tricycle) was added in 1949; two-wheeled tricycle and wide front ends appeared at the same time. MTs also boasted dual hydraulics, which allowed separate lift controls for the left and right side. And a new system of attaching implements greatly decreased the amount of time it took to connect and disconnect them.

A 101-cubic-inch engine delivered about 14 drawbar horsepower and just under 20 at the belt pulley. Although the one-plow-rated tractor was well-suited to smaller farms, it remained in the line only until 1952.

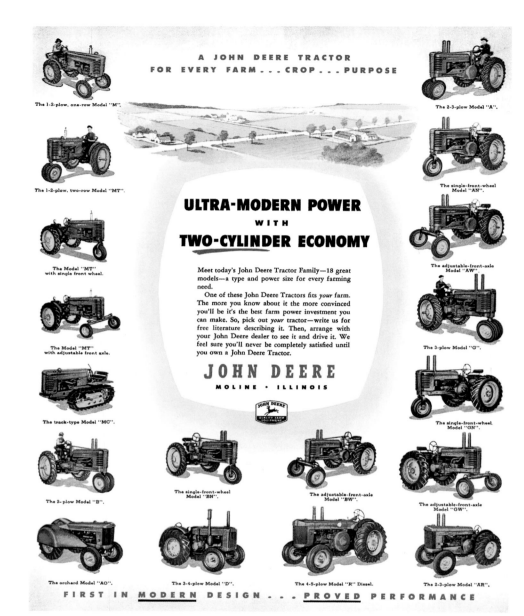

1. The M's vertical inline engine necessitated a driveshaft to bring the power back to the transmission. The clutch was foot-operated. These features had been pioneered on Deere tractors by the diminutive garden-type Models L and LA, which also had vertical, inline two-cylinder engines. The Model M and its successors were built in Deere's Dubuque plant. The M's configuration didn't satisfy all of Deere's customers, however, so in traditional Deere fashion another machine, the MT, was added to the line in 1949. The MT was essentially the same tractor as the M, but could be equipped with an adjustable wide-front, dual- or single-front tricycle wheel.

2. The Model M was Deere's first postwar design and would replace the L, LH, BR, BO, and H models in one fell swoop. MT models like the 1950 example seen here were most often in the three-wheeled "tricycle" configuration. The MA and MT had a telescoping steering column that permitted the driver to stand as he worked. Not the safest way to go about one's business, but it did allow a much more expansive view of the field. A new creature comfort was a backrest that could be inflated to the operator's desire with an internal air bladder. The M series was built between 1947 and 1952, and did well against competing tractors from Case, Ford, Allis, and Farmall.

MODEL G

John Deere introduced the Model G in 1937 as its largest row-crop tractor. Weighing in at 4,400 lb, it was powered by a 413-cubic-inch two-cylinder engine that provided 36 horsepower at the pulley. Early versions of the Model G were unstyled—that is, with no sheet metal covering the radiator or steering gear—and the G was among the last in the line to get the styled treatment.

Styled versions, introduced in 1942, were briefly called the Model GM; this because the government dictated a "price freeze" during World War II, and the new sheet metal kicked up the cost. Deere got around the policy by changing the model name, but reverted back to "Model G" after the war.

Like John Deere's smaller Model A row-crop tractor, the Model G's grille had eight slots on each side of the central steering-gear cover. However, whether styled or unstyled, the Model G had its air-cleaner stack and exhaust stack opposite each other on either side of the hood (on the Model A, they were behind one another in the center of the hood), making them easy to spot. Also like the Model A, the G was described in advertising as a "3-plow" tractor, but those plows could be 16-inchers rather than 14s.

As were Deere's other "Letter Series" tractors, the Model G was redesigned for 1953 and given a numeric designation; in this case, the Model 70. It then evolved into the 720, and later, the beloved 730.

1. The 1937 arrival of the big John Deere Model G was appreciated by the larger-acreage farmer who needed at least some row-crop capabilities.

2. The unstyled and GM versions were available only as dual-front tricycle types. Postwar versions were also available in single front wheel, wide front, and Hi-Crop arrangements.

1. The G was one of the last existing Deere lines to receive Dreyfuss styling, in 1942.

2. The G was Deere's attempt to build a tractor as powerful as the D, but which carried fewer pounds. The new tractor succeeded by creating nearly identical horsepower while saving half a ton in weight. An all-fuel machine, the G ran economically on distillates, but produced more power with gasoline. The G was such a popular model that it was produced until 1953, long after the A and B had been culled from the Deere lineup.

JOHN DEERE
Model "M"
GENERAL-PURPOSE
Tractor

JOHN DEERE
Touch-o-matic
HYDRAULIC CONTROL

JOHN DEERE
QUIK-TATCH
WORKING EQUIPMENT

READY FOR YOU ...with the Right Equipment for your particular needs

"LA"   "L"

3 John Deere Tractors for Small-Acreage Farms

Make this *your* year to put your farm on a new earning basis with a complete John Deere quality-built power farming outfit. You can get the outstanding advantages of John Deere low-cost power in the power size that will exactly fit your needs. The Model "LA" easily handles a 16-inch plow or double-action 5-foot disk harrow, cultivates one row. The Model "L" romps along with a 12-inch plow or a 6-foot single-action harrow, cultivates one row. The Model "H", shown at left, is the John Deere 1-2-plow tricycle-type tractor that cultivates two rows. All three tractors are available with belt pulley and electric starting and lighting; the "LA" and "H" can be furnished with power take-off. Investigate now.

Model "H"

Mail Coupon for Full Information

JOHN DEERE *2-Cylinder* TRACTORS
FOR ECONOMY·SIMPLICITY·EASE
OF HANDLING·DEPENDABILITY

John Deere, Dept. LG-5
Moline, Ill.
Please send me fully illustrated
literature, including equipment,
on tractors I have checked:
☐ 1-Plow, 1-Row "L" and "LA"
☐ 1-2-Plow, 2-Row Model "H"
Name..........................
Town..........................
State.................... R.F.D.....

MODEL 40

What appeared in 1947 as the Model M evolved into the Model 40 when redesigned for 1953. Most other models in the line were also new, and all adopted numeric designations. Revised styling incorporated vertical-slot grilles that mimicked that of the big Model R introduced in 1949. Those with a tricycle front end retained a sheet metal "spine" running down the center of the grille.

With the demise of the Model L and LA in 1946, the Model M had become John Deere's smallest offering, and the 40 carried on that role. Retained on the 40 was the M's vertical two-cylinder engine that defied John Deere's tradition of horizontal twins. Though still sized at 101 cubic inches, output increased by about four horsepower to 22 at the drawbar, 24 at the pulley/PTO.

Like the Model M, the 40 was of "unit design," with the engine and transmission cases serving as the frame to which the front and rear ends were attached. Revised styling made the 40 look huskier than its M predecessor, more in keeping with its 4,000-lb heft.

New was a padded seat to replace the former steel seat pan. Deere's smallest tractor came in a wide variety of configurations and offered numerous features. Tricycle, standard, and low-slung utility front ends were available, along with a crawler version. A three-point hitch was added, as was Touch-O-Matic hydraulics, all of which made the Model 40 surprisingly versatile.

Just What You've Been Wanting—

for ALL-AROUND WORK in FIELDS, ORCHARDS, BERRIES, and VINEYARDS

The Handy, Economical
JOHN DEERE "40"
2-PLOW UTILITY TRACTOR

On the spraying job and other orchard work the "40" Utility, with its low design, snugs right up to the trees for good, close work. Its outstanding fuel economy means many dollars saved each season of its long life.

WHETHER you farm a large or small acreage, be sure to see the "40" Utility, latest of the John Deere two-plow tractors. It's the tractor for you if you have been looking for an economical, low-built tractor for complete power or for helper power. Here is traditional John Deere Tractor simplicity, economy, and dependability in a modern, low-clearance sure-footed power unit with the proper speeds, wheel treads, and tire equipment for your most specialized needs. Standard 3-point hitch for a wide variety of working tools, famous Touch-o-matic hydraulic control, and exclusive Load-and-Depth Control are just a few of its built-in features.

JOHN DEERE
QUALITY FARM EQUIPMENT
JOHN DEERE
MOLINE, ILLINOIS

Send for FREE Literature

JOHN DEERE ● Moline, Ill. ● Dept. J-42

Please send free literature on:

☐ "40" Tricycle ☐ "40" Standard
☐ "40" Utility ☐ "40" Crawler

Name_____
R.R._____ Box_____
Town_____ State_____

Choose the "40" that Fits Your Needs Exactly

TRICYCLE　STANDARD　UTILITY　CRAWLER

Ask your John Deere Dealer for a Free Demonstration

MARCH, 1955

11

THANKS *for the Grand Reception!*

JOHN DEERE "40" Tricycle-Type

JOHN DEERE "40" CRAWLER

JOHN DEERE "40" STANDARD

JOHN DEERE "40" SERIES 2-PLOW TRACTORS

JOHN DEERE
QUALITY FARM EQUIPMENT

JOHN DEERE
Moline, Ill.

We don't deny that a lot of it was expected—the enthusiasm that greeted the introduction of the new John Deere "40" Series Tractors.

After years spent in engineering, building, and testing, we knew we had a series of new tractors that would catch the eye and win the nod of dealers and farmers throughout the country. But we hardly expected the great acclaim that actually has greeted the "40's."

And it's been the same everywhere —a spontaneous, enthusiastic reception that has made all who had anything to do with building these new tractors feel good 'way down inside— that offers continued testimony to an age-old John Deere ax-

iom, first uttered by John Deere himself when he said, "I will not put my name on an implement that hasn't in it the best that's in me."

So we are taking a little time out here to say, "Thanks." Thanks to the John Deere dealers who demonstrated this enthusiasm at introductory meetings at Oklahoma City, Okla., Columbia, S. C., Bakersfield, Calif., Moline, Ill., Minneapolis, Minn., Portland, Ore., Spokane, Wash., and other points— thanks to the thousands of farmers throughout the country for their generous acclaim—and, finally, thanks to everyone . . . for the grand reception accorded these great new John Deere "40" Series Tractors.

JOHN DEERE

Model 40   71

MODELS 50/60

John Deere's popular Model A was redesigned and given the Model 60 designation for 1953. Grille slots switched from horizontal to vertical, and only tricycle versions retained a center sheet metal cover for the steering post. Other easily recognized changes included a new frame design that increased overall length, and a padded seat to replace the steel seat pan.

Further revisions appeared on the mechanical side. A six-speed transmission replaced a four-speed, and dual carburetors helped boost horsepower a bit, now rated at nearly 37 horsepower at the drawbar, and 41 at the pulley/PTO. Also offered was a "live" PTO run off the engine rather than the transmission, and Quick-Change, which allowed faster adjustment of rear tread width. Row-crop versions were available with Roll-O-Matic front-end suspension, which forced one front wheel down when the other encountered a bump, thereby helping to maintain stability, and later models were offered with power steering and a three-point hitch.

Unlike its vaunted Model A predecessor, the Model 60 didn't last long on the market—though that wasn't due to any failings of the tractor itself. Another round of revisions soon prompted another change in designations, and the Model 60 evolved into the Model 620 for 1957.

1. Like the 60, the Model 50 of 1952–56 had Deere "Power Steering" (labeled at the side of the radiator, below the model number).

1.Deere produced the Model 60 from 1952 to 1956, eventually offering it in four iterations: standard, Hi-Crop, orchard, and row-crop.

MODEL 70

A successor to the Model G appeared in 1953, and like other models in the John Deere line at the time, it carried a numeric designation: the Model 70. In keeping with John Deere's commitment to giving the buyer choices, the 70 was available with standard-tread or row-crop front ends. Early models were offered in versions that ran on gasoline, all-fuel (which was similar to gasoline but cheaper), or liquid propane gas (LPG).

All these engines used a new two-barrel carburetor, and the LPG version had aluminum pistons and a stronger crankshaft to handle the engine's higher compression. For 1954, a diesel-powered 70 was added to the roster. Regardless of the fuel, the 70 delivered almost 20 percent more horsepower than the Model G.

Unlike other Deeres of the day, the Model 70's six-speed transmission was shifted with two levers rather than just one. Power steering was now optional, and versions so equipped could be identified by their three-spoke steering wheel (others had a four-spoke wheel) and "Power Steering" decal on the side of the grille. Also located there was the model identification (in this case, "70"), which was helpful in that some of the early number-series Deeres were otherwise difficult to tell apart. All of the early (two digit) number-series tractors had a rather short life. For 1957, the 70 would be replaced by the updated 720.

BIG POWER that Works for "Peanuts"

JOHN DEERE "R" Diesel and Model "70" TRACTORS

Whether you grow grain or row crops on a large scale, John Deere offers you the *big power* you need to make your operations more profitable. It's power that's amazingly economical — that will do more work on a given amount of fuel — that will cost far less to maintain through the years.

Until you've tried a John Deere Model "R" *Diesel* and measured its *unequalled* fuel economy, you've no idea how much better, easier, and more profitable grain farming can be. Here's husky *two-cylinder* power that handles a 4- or 5-bottom plow and similar big-capacity equipment in practically any condition . . . five modern speeds, "live" power shaft and hydraulic Powr-trol, automotive-type steering, plus the kind of operating economy that can cut several hundred dollars off your annual fuel bill.

What the "R" is to the grain grower, the new John Deere Model "70" is to the large row-crop farmer. It's the 4-5-plow tractor specifically designed to handle every tillage, planting, cultivating, haying, and harvesting job easier, faster, better than ever before. In addition to gasoline and tractor fuel, the "70" can be equipped *at the factory* for LP-Gas. It's available with a "live" power shaft, offers "live" Powr-trol and many, many other features as regular equipment.

Your John Deere dealer has all the facts on the powerful, economical "R" Diesel and Model "70" Tractors. See him without delay.

Send for Free Literature
JOHN DEERE

1. The Model 70 was Deere's top-end row-crop machine and produced more than 50 horsepower when it was introduced in 1953. (I-H's Farmall Super M, introduced the year before, had only 44 horsepower.) Besides, the entire Deere line could now be equipped with the Powr-Trol hydraulic pump, Deere's answer to Ford-Ferguson's weight-transferring three-point hitch.

1. Various iterations of the Model D ran on gas, LPG (liquid petroleum gas), diesel, and all-fuel. Whatever the engine, all were two-cylinders. Though introduced in 1953, the popular 70 was produced through 1956, and was good enough to keep customers' minds off the delay of the introduction of Deere's New Generation. The 70 had better hydraulics than that of its predecessor, the Model G, as well as a 12-volt electrical system and power steering.

A **SIZE** for Every Power Requirement • An **ENGINE** for Every Fuel • **EQUIPMENT** for Every Need • at a **PRICE** for Every Pocketbook

Model "70." 4-5 plow row-crop power. Gasoline, Diesel, LP-Gas, or All-Fuel engine.

Interchangeable front-wheel assemblies adapt the Models "50," "60," and "70" to special jobs.

"60" and "70" Hi-Crops meet special clearance needs for growing tall crops.

Standard Model "60" is ideal 3-4 plow power for grain or rice work. Gasoline, LP-Gas, All-Fuel engine.

Model "40" Utility is a lower, longer 2-plow tractor for mowing, hauling, orchard-vineyard work, and many other special jobs.

Model "40" Crawler is the "little giant" on tracks, with amazing stability, traction and flotation.

Hi-Crop Model "40" has 2-plow power and extra clearance for tall-growing crops.

Standard Model "70" for grain or rice growers. Gasoline, Diesel, LP-Gas, or All-Fuel engine.

Moline, Illinois

Quality Farm Equipment Since 1837

MODEL 330

Acting as John Deere's "entry level" tractor in the late 1950s was the Model 330, a close descendent of the Model 320 that had been introduced for 1956. Both models featured a vertical two-cylinder engine (as did the larger 420/430), a remnant of the Model M of the late 1940s. Offered only in a gasoline-fueled version, it displaced 100.5 cubic inches and produced about 21 horsepower. Its four-speed transmission allowed a top speed of 12 mph.

Tipping the scales at less than 3,000 lb and priced just over $2,000, the 330 was often used as a small utility tractor. But the ad at right proclaims it a "1-2 plow tractor," so it could also do conventional farm work, making it the perfect choice for smaller acreages. Like all of John Deere's 30-series tractors, the 330 featured an angled steering wheel, which most farmers found more comfortable than the former vertical placement.

Despite low sales numbers compared with other Deeres of the period—or perhaps because of them—the 330 has since become a prized tractor in collector circles.

MODEL 420

For 1956, the Model 40 was updated to become the Model 420, appearing in late 1955 as the first of the "three digit" Deeres. (Most of its siblings didn't arrive until mid-1956.) Unlike the 40, the 420 would not be the "baby" of the John Deere family.

Due to a 20 percent boost in horsepower—largely the result of a displacement increase for its vertical two-cylinder engine to 113.3 cubic inches—a space opened up for a lower-powered model. John Deere filled the gap with the Model 320, which in many ways could be considered the successor to the Model M, last sold in the early 1950s.

The Model 420 was offered with narrow and wide front ends, along with a tracked crawler version that replaced the former Lindeman-built crawlers. Engines ran on gasoline or all-fuel, and an LPG version appeared for 1958, the 420's final year.

The first 420s were painted all green; later versions added yellow trim as seen on the opposite page. New were an optional five-speed transmission and "live" (engine driven) PTO. For 1958, the steering wheels of most versions were mounted at an angle, which many operators found more convenient than the vertical mounting of earlier models.

THE *NEW* JOHN DEERE *420* TRACTORS

GENERAL-PURPOSE
2-3-PLOW

JOHN DEERE
QUALITY FARM
EQUIPMENT

TRICYCLE AND
STANDARD TYPES

1. The 420 line arrived in 1955 with an all-green color scheme. By June 1956 the tractors had adopted yellow panels that ran along both sides of the hood and at the sides of the radiator. Note that this version does not have an angled steering wheel.

THE New JOHN DEERE "420" FLEET

Brings All That's Good from the Past... plus New POWER and PERFORMANCE

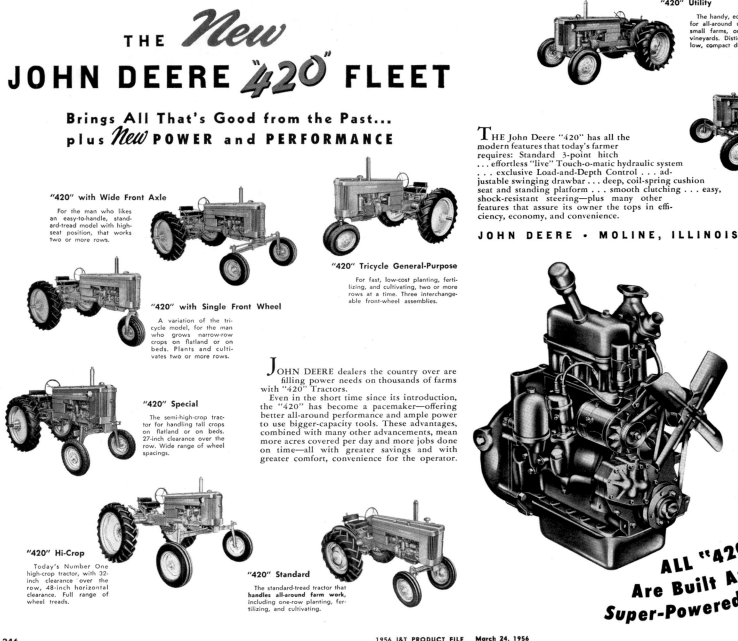

"420" with Wide Front Axle

For the man who likes an easy-to-handle, standard-tread model with high-seat position, that works two or more rows.

"420" with Single Front Wheel

A variation of the tricycle model, for the man who grows narrow-row crops on flatland or on beds. Plants and cultivates two or more rows.

"420" Tricycle General-Purpose

For fast, low-cost planting, fertilizing, and cultivating, two or more rows at a time. Three interchangeable front-wheel assemblies.

"420" Special

The semi-high-crop tractor for handling tall crops on flatland or on beds. 27-inch clearance over the row. Wide range of wheel spacings.

"420" Hi-Crop

Today's Number One high-crop tractor, with 32-inch clearance over the row, 48-inch horizontal clearance. Full range of wheel treads.

"420" Standard

The standard-tread tractor that **handles all-around farm work**, including one-row planting, fertilizing, and cultivating.

JOHN DEERE dealers the country over are filling power needs on thousands of farms with "420" Tractors.

Even in the short time since its introduction, the "420" has become a pacemaker—offering better all-around performance and ample power to use bigger-capacity tools. These advantages, combined with many other advancements, mean more acres covered per day and more jobs done on time—all with greater savings and with greater comfort, convenience for the operator.

"420" Utility

The handy, economical tractor for all-around use on large or small farms, orchards, groves, vineyards. Distinguished by its low, compact design.

"420" Two-Row Utility

The low, wide tractor that straddles and cultivates two rows. Excellent view of work. Big daily capacity. Full range of wheel treads.

THE John Deere "420" has all the modern features that today's farmer requires: Standard 3-point hitch ... effortless "live" Touch-o-matic hydraulic system ... exclusive Load-and-Depth Control ... adjustable swinging drawbar ... deep, coil-spring cushion seat and standing platform ... smooth clutching ... easy, shock-resistant steering—plus many other features that assure its owner the tops in efficiency, economy, and convenience.

JOHN DEERE • MOLINE, ILLINOIS

"420" Crawler (4-Roller)

Brings you all the advantages of track-type power at low initial cost. Delivers 3-4 plow power. Wide variety of agricultural and industrial equipment.

"420" Crawler (5-Roller)

Same power and same general features as the 4-roller model. Extra-long tracks provide maximum flotation, traction, and fore-and-aft stability.

ALL "420" TRACTORS Are Built Around This New Super-Powered JOHN DEERE ENGINE

1956 I&T PRODUCT FILE   March 24, 1956

MODEL 430

As was the case for most other models in the 30-series line, the Model 430 looked little different than its 20-series predecessor, the Model 420. Also like the 420, the 430 was sold for just a couple of model years. Distinguishing the two were the 430's wider yellow trim band on the hood, Float Ride seat, and black-painted dashboard, and some versions carried revised rear fenders.

The 430 was offered in seven variations, including a tracked crawler, making it a very versatile tractor. It shared much with the smaller Model 330, but featured a larger, more powerful 113-cubic-inch (vs. the 100-cubic-inch) vertical two-cylinder engine that came in versions that ran on gasoline, all-fuel, or liquid propane gas (LPG). Since it had more horsepower and was available in more variations than its smaller sibling while being priced just a bit higher, the 430 far outsold the 330.

The 430 forfeits top billing to its Model 530 stable mate, which was powered by John Deere's traditional horizontal two-cylinder engine, but it shows both pulling a three-bottom plow. The caption describing the 430 proclaims, "You've never seen a handier, more efficient plowing team than this John Deere 430 Row-Crop Utility with matched, 3-bottom, integral plow." This was a more glowing assessment than was given to its 530 sibling.

The "430" Row-Crop Utility and the KBL Lift-Type Disk

BIG in everything but costs...

JOHN DEERE 430 General-Purpose Tractors

Big in power, versatility, modern features, ease of operation and dependability . . . but small in cost . . . the John Deere 2-3 plow "430" Tricycle and Row-Crop Utility save time, effort and money on every farm operation.

These modern tractors will boost your farming profits through triple thrift that lets you save on first cost, fuel costs and maintenance expenses. At the same time, you enjoy such modern "take it easy" features as 3-point hitch with exclusive Load-and-Depth Control, Advanced Power Steering, Dual Touch-o-matic hydraulic control, continuous-running PTO, and Float-Ride Seat.

See and drive these tractors that are big in everything but costs—the John Deere "430" Tricycle and Row-Crop Utility.

Ask your dealer for a demonstration soon!

JOHN DEERE
"WHEREVER CROPS GROW, THERE'S A GROWING DEMAND FOR JOHN DEERE FARM EQUIPMENT"

SEND FOR FREE LITERATURE

JOHN DEERE • MOLINE, ILLINOIS • DEPT. X10

Please send further information on the ☐ "430" General-Purpose Tractors ☐ "435" Diesel ☐ KBL Disk Harrow ☐ 127 Gyramor ☐ John Deere Credit Plan.

Name_____ ☐ Student

Rural Route_____ Box_____

Town_____

State_____

NEW 435 DIESEL

Especially designed for new speed and thrift with drawn, 3-point-hitch and PTO tools, the "435" Diesel offers 12 per cent faster speeds than "430" models. Powered by a 2-cylinder, electric-starting General Motors Diesel engine, the "435" readily pulls three 16-inch plow bottoms in most soils. Here it powers the 127 Gyramor.

NOVEMBER 1959 2ad

23

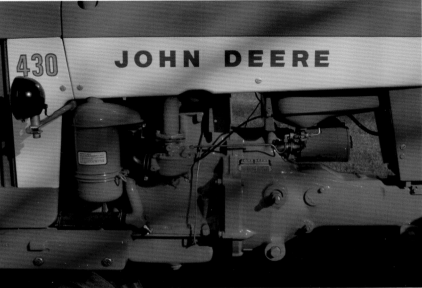

MODEL 630

As was the case with John Deere's other 30-series tractors, the 630 was more of an improved version of its predecessor—in this case, the 620—than an all new design. The 303-cubic-inch engine carried over in a choice of gasoline and all-fuel versions, and an LPG-fueled variant with a cylindrical fuel tank was also offered, all with ratings of about 50 horsepower. However, a new oval muffler helped quiet the distinctive exhaust note.

Like its 30-series siblings, the 630 featured a new paint scheme with more prominent yellow side trim. The steering wheel and dashboard angled upward for greater convenience, and the latter held a starter button to replace the floor pedal used previously. Flat-top fenders could hold four headlights and a radio; rounded "clamshell" fenders were optional. Other options included a Float Ride seat that could be fitted with padded armrests.

The 630 was built for only two years, and this 1960 example closed out the line. It would prove to be the last of the legendary two-cylinder John Deeres, and thus the final edition of the beloved "Johnny Poppers."

Get *Double Satisfaction* at Silo-Filling Time....

The John Deere "630" Tractor and No. 6 Forage Harvester hum through this heavy corn crop in a hurry.

Choose a JOHN DEERE *Tractor* <u>and</u> *Forage Harvester*

YOUR forage acreage, crops, conditions, and power —all are vital factors in getting the maximum return per dollar invested. That's why a versatile new John Deere Forage Harvester and power-matched John Deere Tractor make an unbeatable combination.

On average acreages, the new John Deere No. 6 PTO Forage Harvester handles every crop at rock-bottom costs. A heavy-duty or low-cost Row-Crop Unit . . . 4- or 5-foot Mower-Bar Unit . . . and an efficient Windrow Pickup match your investment to your operation.

For Larger Acreages

On larger acreages, the new John Deere 12 Forage Harvester is your best buy. It features tremendous capacity in both the 6- and 7-foot Mower-Bar Units . . . a heavy-duty Row-Crop Unit . . . and Windrow Pickup.

John Deere Tractors—the "530," "630," and "730" —put the finishing touches on your forage harvesting by providing low-cost power in every crop and condition. Their six-speed transmission . . . Independent PTO

(540 or 1000 rpm) . . . Advanced Power Steering . . . and a host of other features speed harvesting and make work much easier. See your John Deere dealer.

JOHN DEERE

"WHEREVER CROPS GROW, THERE'S A GROWING DEMAND FOR JOHN DEERE FARM EQUIPMENT"

1. In 1959–60, a new Model 630 cost the owner $3,300. The tractor had six forward speeds and produced 48.7 horsepower at the PTO. This row-crop variant has the familiar 30 Series "Oval Tone" muffler.

MODEL 720

John Deere's 20-series tractors that arrived for 1957 offered buyers a wide range of models, but most were similar to those they replaced. The 720 took over for the Model 70 in the line.

A new engine in the 720 produced 20 percent more horsepower than that in the Model 70 for a total of 59. An LPG-fueled version was offered, as shown here; the telltale sign is the huge barrel-shaped tank just ahead of the steering wheel.

A single shift lever replaced the dual-stick arrangement of the 70, and first gear on the 720 was now considered a "creeper" gear. Optional gearing available for 1958 increased top speed to 8 mph. Other optional equipment for the 720 included power steering and Float Ride seat with foam rubber cushion.

A vertical intake stack and rear-facing exhaust were two more choices, and custom Powr-Trol now included a true three-point hitch, along with the capability of powering two remote hydraulic cylinders. As before, narrow-tread versions offered the Roll-O-Matic front end, which forced one wheel down when the other went over a bump, aiding stability and helping to smooth the ride.

Despite its popularity, the 720 was built in 1957 and 1958 only. In late 1958, it was replaced by an equally short-lived 30 series tractor, the 730.

MODEL 730

Most John Deere fans would agree that the Model 730 was among the company's most notable offerings. Not only is it considered a "working collectible" today, but it enjoyed tremendous popularity when new, and ended up surviving well past the time John Deere's other two-cylinder models were put to rest.

The 730 was a versatile workhorse, and on many farms, still is today. It was sold with standard, row-crop, and single-wheel front ends, along with a Hi-Crop version. It was John Deere's largest offering save for the 820, which came only with a standard front end. Its 375.6-cubic-inch horizontal two-cylinder engine came in versions that could run on gasoline, all-fuel, or liquid propane gas (LPG), and a larger 375.6-cubic-inch diesel was also available. Like others in the 30-series line, the 730 carried a different paint scheme than its 20-series predecessor, the 720, with a wider yellow trim band underlining the hood. The steering wheel was set at a more convenient angle, as was the dashboard face.

While production of other models in the 30-series line was discontinued in February 1960, the 730 carried on into the summer of that year, and those destined for export to other countries were built until the spring of 1961. After that, parts were shipped to Brazil, where production continued into 1968.

JOHN DEERE "30" SERIES TRACTORS
offer forward-looking farmers a better tomorrow...today!

It's a big day on any farm when a new John Deere "30" Series Tractor is delivered. To the profit-minded, forward-looking farmer, delivery day represents the arrival of a better tomorrow.

The scene at the right has been—and will be—duplicated thousands of times across farming America. Farmer acceptance of John Deere Tractors has never been higher . . . nor for so many good reasons. Here are quality-built tractors that excel in dependable, low-cost power . . . that are unique in their ability to hold operating costs down. They're unmatched in modern time- and labor-saving features, offering *Advanced Power Steering,* a powerful multi-purpose hydraulic system, versatile 3-point hitch with exclusive Load-and-Depth Control, "live" power take-off and the comfortable *Float-Ride* Seat. On the basis of these facts alone, it's little wonder these tractors have become known as the key to more efficient farming methods and greater profits.

This modern, versatile line of John Deere Tractors is just one more reason why the John Deere franchise is the most valued in the farm equipment field.

JOHN DEERE
MOLINE, ILLINOIS

"Wherever crops grow, there's a growing demand for John Deere Farm Equipment"

For More Details Circle (5) on Reply Card

1. The 730 was another machine designed by Deere to cover the gap until the New Generation units were ready for sale in the summer of 1960. Sold in row-crop, single front wheel, Hi-Crop, and standard tread versions, the 730 was versatile but had only six forward speeds and a solitary reverse. Most 730 production was in Waterloo, Iowa, but some of the tractors were sent in kit form to Rosario, Argentina, and Monterrey, Mexico, for final assembly. All of the Argentinean models were electric-start diesels, while most Monterrey versions were row-crop diesels.

MODEL 820

Topping John Deere's 20-series model line was the mighty 820. It evolved from the Model 80, and except for its yellow hood stripes and wider fenders, looked nearly identical to its predecessor.

Weighing in at nearly four tons, the 820 was powered by a 471-cubic-inch two-cylinder diesel started by a gasoline-fueled "pup" motor. At first the diesel produced about 68 horsepower, but later 1958 models received modifications that boosted the figure to a whopping 75.6 horsepower. These later versions also sported a black dashboard, and are thus known to collectors as "Black-dash 820s"; earlier models, such as the one pictured, had a body-colored dash.

In both cases, the dashboard was beginning to look very carlike, with gauges spread horizontally across a wide panel. With its six-speed transmission placed in high gear, the 820 could sprint along at more than 12 mph, but its intended use was to pull up to six plows through the toughest soil. It was succeeded after only two years by the nearly identical 830, and these tractors represented the pinnacle of John Deere's "two cylinder" era.

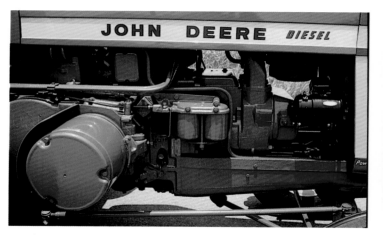

1. Model 820s from early in the 1958 run had all-green dashes; fascias of later '58s were black. Power steering was standard.

2. Its graceful fenders aside, the 1958 Model 820 was a tough puller with a 472-cubic-inch diesel that needed a small, gas-fired "pup" motor to get started.

2000 SERIES

On August 29, 1960, the New Generation John Deere tractors were unveiled with great hoopla at the Coliseum in Dallas, Texas, at an event dubbed Deere Day in Dallas. Deere flew in 6,000 dealers, plus press people, in more than 100 airplanes from all across the country. Fireworks, barbecues, and big-name entertainers supported displays of 136 new tractors and 324 other pieces of equipment.

The new line of Dreyfuss-styled multi-cylinder tractors was an unqualified success. The dealers liked them and the farmers bought them. Sales for 1961 and 1962 were up dramatically. The new line had four models: The 1010 at 30 drawbar horsepower; the 2010 at 40 drawbar horsepower (the 1010 and 2010 were built at Deere's Dubuque facility); the 3010 at 55 drawbar horsepower; and the 4010 at 80 drawbar horsepower. Most were available with gasoline, diesel, and LPG (liquid petroleum gas) engines. They also were available in a variety of configurations, from utility to row-crop.

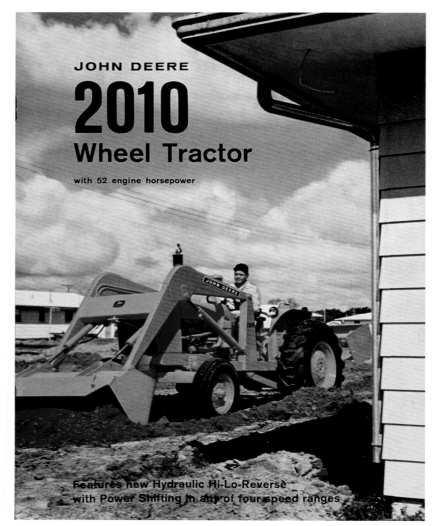

JOHN DEERE

2010

Wheel Tractor

with 52 engine horsepower

Features new Hydraulic Hi-Lo-Reverse with Power Shifting in any of four speed ranges

1. Manufacturing plants in Europe have given Deere a global reach for decades, and produced models not commonly seen in the United States. One of those was the 2120, with engines and transmissions (eight forward gears and four reverse) manufactured in Mannheim and Baden-Wurttemberg, Germany, for use in tractors sold across Mexico beginning in 1971. Mexican law mandated that 60 percent of the tractors' materials be supplied locally, so the 2120 was a true American-European-Mexican hybrid. It ran with a four-cylinder engine that was given an added boost with turbocharging.

1. The "Old Style" 2640 was introduced in 1975, as the latest Deere series that stretched back to 1962. A six-cylinder engine produced 80 horsepower that the operator could put to work via six forward gears. Within the series, the 2040 and 2240 were imported from Deere's Mannheim, Germany, plant while the 2440 and 2640 hailed from Dubuque, Iowa. Any of the four could be ordered in Orchard or Vineyard trim.

2. In the early 1960s, several of the leading agricultural equipment companies created standardized products to be built by their international operations. In 1963, Deere & Company followed suit. Two years later, three new "Worldwide" tractors were introduced. This trend has been expanded since then to include the entire tractor line. Deere's "worldwide tractor" project was unveiled in 1965 with the Model 2020. Because the 2020s were intended for use in America and abroad, several variations were available. The four-cylinder 2020 was built in Dubuque until 1967, when assembly shifted to Deere's Mannheim, Germany, and Saltillo, Mexico, facilities. This 2020 is sheathed in an orchard package that encloses the hood and rear wheels to protect trees during operation. Even the operator's pillion has been cloaked in a protective screen that allows forward visibility without danger to driver or trees.

3. The 2510 of 1965–68 was designed for farmers whose needs fell between the capabilities of Deere's 2020 and 3020 models. And indeed, the relationships were very direct: When a 3020 chassis was altered and fitted with a 2020 engine, the 2510 was born. It was sold in gasoline and diesel versions, as well as row-crop and Hi-Crop iterations. Both engines ran with four cylinders, with the gas version displacing 180.4 cubic inches, the diesel 202.7. Buyer's choice of a Synchro-Range or Power Shift gearbox provided eight forward speeds. Base price for a row-crop gas-powered model was about $3,975, with diesel, Hi-Crop, and the Power Shift options adding to the bottom line.

3000 SERIES

The expanding interest in the hobby farming sector, as well as small commercial needs, was the inspiration for the debut of the 3000 Series. The series was sold in four variations, beginning with the introductory 3005—it featured a gear-driven transmission only, and was ideal for constant-speed demands. The 3E models was another medium-frame tractor that was standard with a hydrostatic drive. The 3020s added features and options to meet with the demands of a larger operation. They could also be purchased with a cab, adding more comfort and convenience when the weather was not as hospitable, and could also be had with heat and A/C.

What's more important than gallons per hour?

Acres per Hour!

1. An ad from about 1966 emphasizes the work efficiency of Deere tractors. "More usable horsepower at faster ground travel speeds" was an advantage pointed out in the ad's copy.

2. Evolution being what it is, the tractor world produces many new models that closely mimic their predecessors. The 3020 was a kissing cousin to the 3010 when it came from Deere's Waterloo facility in 1964, but it was more desirable because of additional power, traction lock, and the available Power Shift transmission. For even more convenience, the 3020 gained optional Front Wheel Assist in 1968, and could be had with orchard-friendly sheathing.

1. Yanmar of Japan was founded in 1912 and made its name with diesel engines. It entered the agricultural equipment business in 1961, bringing its diesel technology to a successful line of small, modestly powered tractors that established a presence in the United States. For Deere to design and build from scratch models that were similar to Yanmar's would have been economically infeasible, so in 1977 a co-op deal was struck by which Yanmar would restrict the sales of its branded tractors to overseas markets, while Deere would sell them stateside and in Canada as Deere machines. The 20- to 40-horsepower diesel models came to North America adorned in the familiar green and yellow livery, and sold well.

2. As farmers spent more and more time in their tractors, the need for additional safety and convenience became apparent. By adding an enclosed cabin, as on this 1979 Model 3130, Deere shielded operators from flying debris, dust, and other day-to-day hazards of working a field. The cabin also added structural strength that helped to protect the driver in the event of a rollover. Many modern Deere models equipped with cabins have air conditioning, audio systems, and GPS systems that provide the farmer with comfort and information.

3. After a brief yet excruciating delay, Deere's New Generation tractors made their way onto the market in 1960. The new 3010 was a four-cylinder version of the larger 4010, but delivered on the promise of being a wholly new breed. A new, triple-circuit hydraulic system corrected a long-standing drawback by which all components had had to share the same hydraulic line, which greatly reduced efficiency. The new setup allowed the 3010 to have separate circuits for power steering and power brakes, and a third for the three-point hitch or any other remote implement. When fitted with the LPG (liquid petroleum gas) propane option, the 3010 was the only Deere model that concealed the fuel storage tank beneath the cowl; earlier Deeres made do with a bulbous tank located outside the lines of the long hood.

4000 SERIES

John Deere introduced their 4010 in 1960, and at the time it was one of the largest models the company had offered. Designed for smaller applications, the 4000 Compact Series rolled out in 1998. The new lineup was designed for use on smaller farms and some commercial applications. The series included models ranging from the 4100 to the 4700, and every step up the ladder offered more horsepower and added features. Twenty horsepower was listed on the 4100, and the largest 4700 claimed 48 ponies on hand. Four of the seven models were discontinued in 2001, with the remaining three seeing their end in 2003.

As seen in John Deere's long history, constant improvement of the company's products was paramount. In keeping with that philosophy, the 4000 Series returned to the fold beginning in 2004 with the 4120. In 2005, the 4320, 4520, and 4720 joined the roster, and finally the 4005 compact tractor arrived. Every step in the model-line evolution added more power and features to suit the needs and the demands of a varied group of buyers. Among the 4000 Series, the 4020 models also could be ordered with the ComforGard Cabs. These cabs provided the operator with far better chance to produce during all forms of weather.

In *New Generation* JOHN DEERE ENGINES...

bigger, heavier rotating parts follow through where others stall out

Farming with a New Generation Tractor, you'll move through big jobs without hesitation . . . rarely, if ever, shifting down to work through tough spots. Here's why: traditionally heavy John Deere rotating parts, shown in the "4010" Diesel engine cutaway, build up tremendous energy (a reserve of power) for real lugging ability.

Note the larger pistons, heavier connecting rods, bigger-diameter piston pins, and massive crankshaft (150 lbs.)—backed by a heavy flywheel and clutch assembly. Put this beefy engine assembly into motion and it just "plain and simple" doesn't like to be stopped!

Another thing, John Deere builds this engine to "deliver" dependably over many seasons of hard use. The "4010" crankshaft, for example, is carried on seven main bearings in a solid foundation to "stay put" under the most extreme loads.

These advantages, together with dozens of other advanced features, make a John Deere New Generation Tractor your best buy for efficient, dependable, trouble-free farming power. Your John Deere dealer has a type and power size to fit your needs exactly.

JOHN DEERE • 3300 RIVER DRIVE, MOLINE, ILL.

Arrange for an on-your-farm demonstration of a New Generation 35 h.p. "1010," 45 h.p. "2010," 55 h.p. "3010," or 80 h.p. "4010." Contact your John Deere dealer today. Ask about his liberal John Deere Credit Plan, too.

JOHN DEERE design, dependability, and dealers MAKE THE DIFFERENCE

MODEL 4010

To John Deere traditionalists, the New Generation models introduced in October 1960 came as quite a shock. Many couldn't believe the company would forsake the two-cylinder arrangement that had always been John Deere's trademark—and greatest selling point. But only so much power can be efficiently wrung from two cylinders, and John Deere had maxed out the design's potential. With farmers asking for more, the company needed to head in a new direction.

The first New Generation John Deere to hit the fields was the 4010—destined to be the biggest in the initial lineup. Power came from an inline six-cylinder (gasp!) engine, with gasoline, LPG, and diesel versions offering up to 75 horsepower. Standard and row-crop front ends were available, as was a Hi-Crop variation. Not only was the 4010 more powerful than any of its predecessors, it also introduced some new features, such as an eight-speed transmission, an innovative central hydraulic system, and an orthopedist-designed seat that became a benchmark for comfort.

1. When the 4010 took to the fields for the first time in 1961, it was the biggest non-articulated tractor John Deere offered. One of the New Generation models, it was powered by an inline six-cylinder engine that could be ordered with gas, LPG, or diesel fuel options. The gas and LPG engines displaced 302 cubic inches, while the diesel was listed at 380. More than 75 horsepower was on tap, another first for Deere. An eight-speed Synchro-Range gearbox could move the nearly five-ton tractor along at 14.3 mph. Henry Dreyfuss and Associates designed the cockpit and seating, in conjunction with an orthopedic specialist. The seat was the most comfortable yet installed on a Deere.

2. Resale value of equipment can be critical to farms that regularly upgrade. This ad, from about 1988, declares that Deere is everybody's best bet.

MODEL 4020

Despite some initial misgivings, the New Generation John Deeres introduced in 1960 met with almost universal acclaim. The 4010 had made a name for itself as the top-line model, but Deere soon decided some detail tweaking was in order.

Still fitted with the inline-six first used in the 4010, the 4020 delivered even more power. Gasoline or LPG fed a 340-cubic-inch engine, while the diesel measured 404 cubic inches. Two transmissions were offered, both with eight speeds: the standard Syncro-Range, or the easier-to-use Power Shift, which allowed shifting without using the clutch. In top gear, a 4020 could zip along at nearly 19 mph.

Row-crop, standard, and Hi-Crop versions were offered, with base prices spanning a range of $4,714 to $5,366 in 1965. A protective cab was available as an option. As opposed to its immediate predecessors, the 4020 enjoyed a relatively long production run, the final examples rolling off the line in 1972. Many of these tractors are still in use today, and they often cost more on the used market than they did when new.

1

1. The 4020 diesel was Deere's update of the popular 4010, which was among the first of the New Generation tractors of 1960. This 4020 is a Hi-Crop variant, which marks it as about a 1965 model.

2. As good as the 55-horsepower 4010 was, the 4020 trumped it and became one of John Deere's greatest models ever. The 4010's incredible power was increased, to 94 hp, and the cockpit was redesigned for driver comfort. The buyer now was given the choice of the eight-speed Synchro-Range gearbox or the latest Power Shift. The latter system's single lever was seen as a great convenience, although it did take away a few of the 4020's horsepower in the process. 4020s built for 1969 and beyond had even more features, including a new control panel and an improved 12-volt electrical system with an alternator. It's not difficult to see why 4020 models purchased today cost far more than they did when new.

Put John Deere power in front of a minimum-tillage hookup and you can cut production costs as much as $5 an acre. Makes good farming sense—put more of a load on tractors . . . double-up on tools . . . get several operations done once-over.

John Deere saw the handwriting on the wall in the need for bigger tractors and put out models ideally suited to double-up operations. The 65 h.p. "3020" * and 91 h.p. "4020" * deliver power for multiple-unit hookups not only on the drawbar but through the PTO and hydraulic systems as well.

Hand-in-glove with the development of these tractors, John Deere engineered equipment-ganging devices to make big tractor power pay off. Minimum tillage hitches. Squadron hitches that link together up to 40 feet of drills. Tool carriers brawny enough to handle several types of implement attachments simultaneously.

Your John Deere dealer wants you to take a crack at farming with double-up power. He'll supply the tractor and implements if you'll supply a few hours of your time. When the trial is over, he'll make it easy to shake hands on a tractor and implement deal with good John Deere Credit Plan terms.

Diesel model

John Deere double-up power!

JOHN DEERE
Moline, Illinois

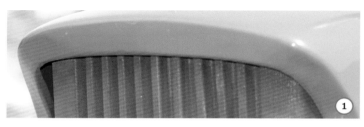

RESEARCH & DEVELOPMENT

Deere has routinely allocated a greater portion of its profits into research and development than most such companies. Technically advanced new products are thoroughly proven before being released to customers. The Deere Research Center, for example, has facilities for testing implements and tires in various kinds of soil. There's also a Cab Simulator that rivals an aircraft flight simulator. It can be programmed to give the operator the exact sensations encountered in operating, for example, a combine on a particular sort of terrain.

1. The 1963–66 Model 4020 was one of the most popular New Generation Deere tractors. Its six-cylinder engine (gasoline, diesel, or LPG) hidden under the Dreyfuss styling was mated to an eight-speed Power Shift transmission; an automatic trans was optional. Tested horsepower at the PTO shaft was 95.8.

2. At the University of Nebraska at Lincoln in 1982, Test #1458, of a Deere Model 4250 tractor (15-speed Power Shift diesel), was carried out at the UNL test track. This and all other UNL tests followed the Agricultural Tractor Test Code approved by the American Society of Agricultural Engineers and the Society of Automotive Engineers. The silver vehicle seen here is a tractor test car that was hooked to, and reported on, the 4250's vital functions. PTO and drawbar power were evaluated; likewise fuel consumption and the tractor's performance during two hours at maximum engine power.

5000 SERIES

The 5000 models were seen in a huge variety of configurations, providing all of the power and features needed for a larger farming operation. Constructed from components delivered from across the globe, assembly for the 5045E through 5075E versions took place in India.

The 5000 E Series machines featured mechanical front-wheel drive for increased traction. Other selections were a choice of the 12F/12R PowrReverser or 9F/3R SyncShuttle transmissions. A three-point hitch was standard and an optional Economy PTO feature saved fuel and reduced noise. The SE models provided additional gear at a "value-minded" price.

The larger 5000 M Series was powered by the PowerTech diesel engines, which gave the owner a more robust machine to suit rigorous needs. The M models could be delivered in the fixed, open-station design or with a climate-controlled cabin. The enhanced power of the M Series was matched with contemporary and environmentally sound equipment for use in today's eco-conscious society.

PRODUCTION OPPORTUNITY '67

New 5020 Row-Crop Tractor...
132 h.p....world's most powerful

MORE OPPORTUNITIES NEXT PAGE

You're equipped to produce a lot more groceries, and show a better net profit, with a new 132 h.p. "5020" Row-Crop. It widens the measure of your ability as it tightens the belt on your production costs.
Think Big—You'll plow with 8 bottoms at a steady 4-1/2 mph . . . finish fields faster, at lower cost, than you could with two 4-plow tractors.

Hook onto an 8-row minimum-tillage rig . . . plant in 30-inch rows . . . cut trips over the field 50 percent. You'll require fewer fair-weather Spring days to complete planting. Cultivate as many rows with a "5020" Row-Crop in one pass as you did in two or three trips before. With a 1,000 rpm forage harvester powered by a new "5020,"

you'll chop—even recut—two heavy-as-they-come rows non-stop.
It's unbeatable—The "5020's" variable-speed 6-cylinder Diesel works hand-in-hand with the 8 overlapping speed ranges of the Syncro-Range Transmission. A speed for every need is supplied—from a 1-1/2 mph crawl for powershaft work to a rapid transport 20 mph. You're in finger-

tip command of hydraulic power for steering, braking, implement control, PTO and differential-lock operation. And the new "5020" Row-Crop has more tractive weight than any other over-100 h.p. row-crop tractor. Yet it treads lightly as it delivers big-tool pulling power to drive wheels. Here's why:
New Rear-Wheel Option: Double

rear wheels, shown above, are individually adjustable on the axle. Set them in for plowing . . . space them to straddle rows for cultivating . . . or drop off the outer wheel for rebedding. Each securely fastened, heavy, cast wheel can transmit full horsepower.
Seize the opportunity . . . field-test a "5020" Row-Crop soon.

MODEL 5020

With the advent of John Deere's New Generation six-cylinder engines, power reached unprecedented levels. The diesel in the mid-1960s 5020 churned out 132 horses from 531.6 cubic inches, enough to lay claim in contemporary ads to being the "world's most powerful."

Initially, the 5020 came only with a standard front end, but a row-crop version was added in 1967, making the big tractor even more versatile. Options included a cab with heater and air conditioning, remote hydraulics with single or double output, and a three-point hitch with Quik-Coupler and live PTO. An eight-speed Syncro-Range transmission provided speeds from 1½ mph to 20 mph, and ads stated the 5020 could "plow with 8 bottoms at a steady 4½ mph"—a valuable asset to many farmers, as they'd need "fewer fair weather spring days to complete planting." Indeed, the 5020 could do twice the work of many large tractors built just ten years before, and in the hands of an able farmer, would become a significant contributor to the food baskets of the world.

1. As farm operations grew larger to meet market demand, farmers' desire for additional power became a roar. Until the mid-1960s 75 horsepower was deemed adequate for most uses, but even that lofty output was about to go by the wayside. Part of the New Generation line, the 5010 set new standards for Deere power when it was introduced in 1966. With a six-cylinder diesel under the cowl and an eight-speed Synchro-Range gearbox, the 5010 churned out 121.12 horsepower at the belt. As Deere's first two-wheel-drive model to exceed 100 horsepower, it was welcomed by farmers with large properties. The 5010 was available in standard tread and diesel only. When new, the tractor carried an MSRP (manufacturer's suggested retail price) of $10,730 and weighed more than 17,000 lb when loaded for work.

1. A six-cylinder diesel displacing 531 cubic inches powered the 1966–72 5010. At the rear, the 5010 had a sophisticated Category 3 three-point linkage. Optimal engine performance came at 2,200 rpm.

MORE POWER FOR YOU—AND THE GUTS TO BACK IT UP

NEW 178-hp AND 255-hp JOHN DEERE SP FORAGE HARVESTERS OFFER HEAVY-DUTY, PROTECTED DRIVES TO MATCH POWER OUTPUT

2

3

The new 5440 and 5460 Self-Propelled Forage Harvesters bring you big power backed by heavy-duty construction and built-in reliability.

The 178-hp diesel on the compact 5440 offers plenty of power to farmers who need to step beyond pull-type capacity in their operation. Consider a further step up to the 255-hp 5460 if your acreage and feeding program demand really big production day after day, year after year in heavy harvests. The big diesel power units in both of these harvesters are John Deere designed, built, and warranted.

But we didn't stop with big engines. You get beefed-up front and rear axles. More strength than ever in the cutterhead housing, augers, and feed rolls. An optional 4-wheel drive.

There's protected design, too, besides brute strength. For example, an electrically controlled clutch on the feed rolls. This clutch takes drive pressures off the gearcase for easy, no-load shifting. And on the 5460, the main drive gearcase is equipped with an oil cooler so you can operate at higher load levels.

Big capacity continues through the giant 24-inch-diameter cutterhead. Nine heavy, tungsten-carbide-faced J-knives cut the material and move it quickly out of the cutterhead. And for clean-cutting efficiency that

lasts, you get reverse knife sharpening and tungsten carbide on the stationary knife vertical edge.

Standing or windrowed hay, whole-plant silage, ground ear corn, stalk residue...the 5440 and 5460 can handle them all.

All this toughness and power doesn't sidestep comfort. The new cab, with large rear window, provides a full field of vision. You'll enjoy the comfort of the John Deere Personal-Posture™ seat and pressurized cab. Or enjoy a cool environment on the hottest days with optional air conditioning.

Two big-power, self-propelled forage harvesters with built-in reliability and comfort are at your John Deere dealer. See him for full details.

THE FORAGE SPECIALIST

1

1. In the late 1970s, Deere introduced a pair of related, self-propelled forage harvesters, the "compact" 5440 (background) and the larger 5460. These were heavy-duty machines designed to cut and chop numberless rows of stalks and other forage into silage suitable as feed for livestock. An oil-cooling system and versatile clutches helped protect the gear cases from burnout. Because the tungsten-carbide cutterheads on both of these machines were fully two feet in diameter, many acres could be cleared in a day. "Personal-Posture" seats, and optional air conditioning and four-wheel drive, made the work a little easier.

2. Personnel at Deere's engineering-driven Technology Center in Pune, India, look closely at ways to reduce the time needed for product development. The results have ramifications for Deere operations all over the world. This factory has been noted for its sterling safety record as well.

3. The Model 5403 is an emerging-markets product manufactured by Deere at its facility in Pune, India. This 74-horsepower machine entered production in 2007.

EXPANDING MANUFACTURING CAPACITY

In the early 1960s, Deere began integrating Mannheim, Germany-based Lanz products into the U.S. tractor line. More and more production of mid-sized tractors of the World Tractor concept (tractors well suited to profitable use in emerging agricultural economies) came from Mannheim and other European plants. By 1987 Deere saw the need to increase production of its smaller tractors, and built a plant for that purpose in Augusta, Georgia.

1. The 5320 was built from 2000 to 2004, and followed closely on the heels of the 5310. Both were built at Deere's Augusta, Georgia, facility. Power was provided by a three cylinder diesel displacing 179 cubic inches. Output of 64 horsepower was coupled to nine forward gears and three reverse. The standard model was two-wheel drive with four-wheel as an available option.

2. Deere's 89-horsepower 5510 was built at Augusta, Georgia.

3. Deere bought the Mannheim, Germany-based Heinrich Lanz Tractor Company in 1956. The Deere factory there grew to become the largest exporter of tractors in Europe—it accounts for more than half of Germany's tractor production. Here, a truck driver steers his load of new Deere tractors out of the company's huge Mannheim works.

1

2

3

6000 SERIES

John Deere's 6000 Series made its debut as a 1993 model, and has been sold in more than seventy variations since then. The 6000 E Series arrived in 2010 with two models, followed by three more in 2016. Between the E and M series, the E was the bigger and more versatile option. Both variations could be delivered with open-station or cab layouts, as well as a number of transmission choices. The 6M was considered to be the best suited for all-around duties at the farm, while the 6E was also adept at a number of chores without regard for their difficulty.

1. In 1992, the first off the assembly line of the new Augusta, Georgia, facility was the "Thousand Series." In 1993, the 5000, 6000, and 7000 Series tractors boosted Deere's market share in the U.S. and Europe. Buyers in Germany, for instance, could choose from among tractors made by 20 companies, yet Deere bolted upward from number three to number one in tractor sales in that nation. Horsepower was one reason.

2. With a trailer full of clippings in tow, this self-propelled 6910 forage harvester from about 1996 is off to the storage facility to complete the silage-cutting process. Even smallish farm operations can generate silage by the ton, which is used not just for livestock forage but for anaerobic digestion, a process by which microorganisms break down the fermenting grass in an oxygen-free environment, creating biogas that can be used for heat and electricity. In addition, the solid, nutrient-rich digestate that remains after anaerobic digestion is a common ingredient of quality fertilizer.

1. A baling machine that labors behind a Deere tractor is a common sight in the fields. The Deere Model 6900 tractor, produced in the Mannheim factory during 1994–97, ran with a 414-cubic-inch turbocharged six developing 130 horsepower, so the British-made Claas Quadrant 1150 baler posed no challenge. The baler's duty is to collect hay from the ground and neatly wind it into a tight bundle. The resulting bales can be easily transported, or can be left in the fields for cattle to feed from.

3. The John Deere plant in Mannheim, Germany, was busy in the 1990s. One of its products was the 6400, which was motivated by a four-cylinder, turbocharged diesel that produced 85 PTO horsepower. The tractor was sold in two- or fourwheel- drive variations, and listed two different gearboxes: the Synchro-Plus, with 12 forward speeds and four reverse, and the PowrQuad, with 16 and 12, respectively. When prepped for action the 6400 weighed just over five tons. Production ran from 1992 to 1998.

2. The Model 6850 is a highly specialized, self-propelled forage harvester. These machines utilize a cutterhead or flywheel to chop the material on the ground into silage before passing it to the onboard storage area. Once passed through the harvester, the silage can be treated with additives that accelerate fermentation. Much like a crop harvester, the 6850 and similar Deere forage harvesters can accommodate hundreds of bushels before having to offload to an alternate wagon or truck.

4. The 6020 series included five four-cylinder models, plus five with bigger six-cylinder motors. Of the four cylinders, the 6420S (shown) was at the head the class. A turbocharger helped produce 120 horsepower that fed through a stepless transmission offering seamless gear changes and almost endless ratios. Premium Plus versions featured AutoPowr that allowed a top speed of 31 mph. Two-wheel drive was standard, with four-wheel an available option. The cockpit of the 6420S looked like a gamer's paradise, with a wealth of electronics and control levers.

7000 SERIES

The big 7000 series stormed onto the scene in 1992 with the 7600, 7700, and 7800. The 7800 was the most powerful of the three, running with a turbocharged, 466-cubic-inch six that was coupled to a PowrQuad transmission. Sixteen forward gears and 12 reverse were on hand, with easy access provided by Deere's Power Shift setup.

When equipped with four-wheel drive the 7800 tipped the scales at 15,560 lb, a capacious weight that included 91 gallons of fuel. The spacious cab could be entered from either side and offered AM/FM radio and other amenities not typically found on farm implements. When the 7800 ceased production in 1996 its MSRP was $84,000.

Here's an inside look at the future of farming. These new tractors are not modified versions of previous John Deere models. They feature virtually 99 percent all-new parts, all-new technology, and an all-new way of accomplishing tasks today ... and tomorrow.

❶ *Discover the all-new TechCenter Cab on pages 4 and 5.*

❷ *See how we improved engine power, performance and reliability on pages 6 through 9.*

❸ *The new 19-speed Power Shift transmission and new PowrQuad transmissions are shown on pages 10 through 13.*

❹ *The all new John Deere designed hydraulic system is detailed on pages 14 and 15.*

❺ *See page 16 for the new John Deere electrohydraulic hitch.*

❻ *New front and rear PTOs are detailed on page 17.*

❼ *All-new steering system is described on page 18.*

❽ *Turn to page 19 for information about the high-performance tractor and trailer braking systems.*

❾ *New John Deere planetary final drives are covered on page 20.*

❿ *Tight-turning Caster/Action MFWD is shown on page 21. You'll find full specifications for new John Deere 7000 Series Tractors on pages 22 and 23.*

⒈

1. Built at the Waterloo, Iowa, plant between 1992 and 1996, the Deere Model 7800 was a popular tractor at home and abroad. The 466-inch turbocharged diesel delivered 145 horsepower to the PTO. PowrQuad was the standard transmission, with 16 forward ratios and 12 reverse. The optional Power Shift setup claimed 19 forward ratios and 12 reverse, but as in all Power Shift versions gear changes could be made without using the clutch. The 7800 weighed a scant 9,000 lb when ready for duty and sold for $84,000 at the end of its production run in 1996. This one pulls a Tyler fertilizer spreader.

8000 SERIES

John Deere upgraded its midsized tractors (those developing 66 to 145 horsepower) in the early 1990s. The very large articulated tractors were also improved, with the largest, the Model 8970, upgraded via a 400-horse Cummins engine. These 8070 tractors had electronic engine control that increased the power output of the engine as it lugged down by boosting the intake pressure and fuel flow. This feature was named "Field Cruise."

For the 1997 tractor lineup, Deere introduced rubber tracks as an option for its 8100 through 8400 Series tractors, which were otherwise four-wheel drive. The power range was 160 to 225 horsepower. Thus, Deere & Company launched its challenge to Caterpillar. By model-year 2000, the Model 8400T had become the 8410 and other models in the 9000 series were also upgraded.

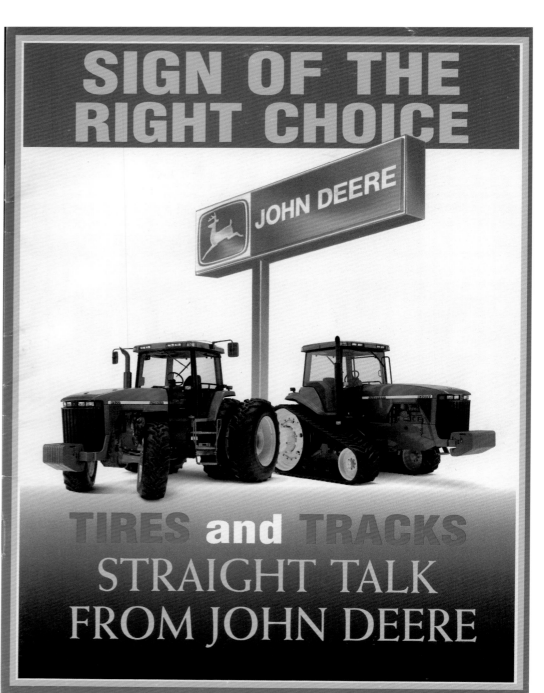

1. Mindful of the ever-increasing demands of modern farms, Deere introduced the 8100 in 1994. Built on a 116.1-inch wheelbase and powered by a six-cylinder turbocharged diesel, the 8100 was nothing to trifle with. The 466-cubic-inch motor delivered 160 PTO horsepower and offered the operator 16 forward speeds and four in reverse. Inside the enclosed cabin you might find air conditioning and an AM/FM/cassette stereo system to provide some comfort while working the soil. Sold in a choice of two- or four-wheel drive, the 8100 weighed nearly 18,000 lb when prepped for use. The tractor could carry 135 gallons of fuel, enough to provide a full day's work from a single fill-up. The 8100 was built through 1998.

2. The John Deere 8010 caused much amazement when it was introduced at Deere & Company's field day in Marshalltown, Iowa, in 1959, because of its huge size: It was 20 feet long, 8 feet wide, and 8 feet tall. It weighed 20,000 lb without ballast and 24,000 lb with liquid in the nearly 6-feet high tires. The 8010 ran with a six-cylinder, two-cycle GM 6–71 "Jimmy" diesel engine of 215 horsepower (no other Deere tractor had yet exceeded 80 hp), and it had a nine-speed transmission when the most any other Deere had was five. Instead of mechanical brakes, like every other Deere, the 8010 had air units.

3. Deere's 8020 articulated tractor was based on the 8010. In fact, many 8020s were rebuilt 8010s that had been recalled by Deere because of serious transmission problems. All 8010s that were recalled and repaired were returned to owners and dealers with "8020" badging. The 8010 design dated to 1959, with the follow-on 8020 showing up in 1963. 8020s had an upfront provision for the mounting of a 'dozer blade, with front-end reinforcement that allowed for the mounting of a blade's lift cylinders. Of the 100 8010/8020 units built, about 75 remain registered with collector groups.

1. The massive, articulated Model 8630—one of Deere's Generation II models with fresh sheet metal—was introduced at an industrial show in Saarbrucken, Germany, in the fall of 1972. This 24,200-lb, four-wheel-drive tractor entered regular production in 1975, and came with Deere's brand-new Sound-Gard cabin as a standard feature. The cabin kept sound levels impressively low, even when the 8630 was grunting at its rated 2,100 rpm. The 16-speed Quad-Range gearbox was another first.

2. In 1982, following on the heels of the diminutive Model 650 (at 1,530 lb, the smallest farm tractor Deere had ever built), the company rolled out its largest to date: the 37,480-lb Model 8850. It was the biggest of a four-wheel-drive series that included the 8450 and 8650; all three were articulated (center-hinged) to allow for tight cornering. The 8850 ran with an eight-cylinder diesel that displaced 955 cubic inches and produced 304 PTO horsepower. Given the machine's immense weight, it's no surprise that it drank 19 gallons of fuel during every hour of PTO operation. Sixteen forward gears and the eight-wheel option seen here made the 8850 a real workhorse.

3. In the world of gigantic tractors, the John Deere 8970 of 1993–96 (a '93 is pictured) was one the biggest. Riding a 143-inch wheelbase, this articulated-chassis machine weighed nearly 34,000 lb when the required 220 gallons of diesel fuel and 72 quarts of coolant were added. An octuplet of massive rubber wheels gave the 8970 plenty of traction regardless of terrain or payload. The 855-cubic-inch turbocharged diesel developed 400 horsepower at the PTO. Deere's Syncro transmission was the standard gearbox, with 12 forward ratios and three reverse. The optional PowrSync setup brought an additional 12 forward gears. In 1993 the 8970's price tag was nearly as big as the tractor itself: $145,000.

It's not the similarities that upset our competitors . . .

1. Deere's 8000T series (*T* for tracked) came on line in 1997. They were durable machines designed to handle any light- to medium-duty agricultural chore.

2. With 255 horses at the PTO, and a weight exceeding 23,000 lb, the Model 8520 isn't fazed by heavy road-hauling chores. The standard rear-lift capacity at the three-point hitch is 15,180 lb; an optional setup brings that figure to 17,300. The operator can make use of 16 forward gears and four reverse to get the most from the six-cylinder diesel engine. This one is fitted with optional driving lamps and extended side mirrors. The 8520 was manufactured at Deere's Waterloo plant from 2002 to 2005.

3. The 8410T, an 8410 variant, was a dominant tracked machine of the early 2000s. Its 496-cid turbo diesel produced 270 horsepower at the PTO. An electronic control system linked to the engine helped prevent inadvertent operator overload of the drivetrain.

4. Deere's Model 8230 dates from 2006 to 2009. It's a hardy row-crop machine that runs with a 265-horse diesel producing 232.5 maximum hp at the PTO. A well equipped 8230 will have front xenon lights, FM business band radio, a 60-gpm (gallons per minute) hydraulic pump, front fenders—and even a dual-beam radar unit to keep precise track of distance and ground speed, area covered, and area per hour. All of that matters because the operation of many modern planters, fertilizer applicators, sprayers, and seeders is controlled by radar speed. In independent analysis, the Nebraska Tractor Test Lab found that the 8230 was nearly 25-percent more fuel efficient than the comparable New Holland TG245 and Case I-H Magnum 254.

5. If Deere has learned anything in its history it's that people buy many sorts of tractors for a boggling array of chores. Here, a John Deere pulls a beach rake to groom sand at a public vacation spot. Similarly, a gang mower is an efficient device to maintain any golf resort's manicured grass. Winter brings a new set of tasks, and Deere rises to the occasion with tractors and other machines designed for basic snow removal or careful grooming of ski slopes.

9000 SERIES

Reigning at the top of the John Deere lineup is the 9000 Series—which debuted in 1997. Since then, more than forty variations have been featured in the lineup. The 9R Series appeared with the first range of nine different versions, with four additional variants in 2016.

At the bottom of the heap we find the 9370R, powered by a 370-horsepower 9.0-liter engine and boasting a long list of power and convenience options to suit the needs of any large-operation farm. The 9620R and RX versions are driven by massive 620-horsepower, 13.5-liter mills and feature an even longer list of offered equipment. Choices in the PTO and three-point hitch segments provide all the options required to build a tractor to satisfy most every need.

All of the powertrain choices are mated with a lengthy list of creature comforts to ensure the performance of the farmer while on duty. The CommandView III Cab combined with Deere's CommandArm and CommandCenter provide a high-grade sound system, Bluetooth device connectivity, complete heat and A/C controls, and satellite radio.

Are you ready for the Big Time?

Get ready for the most powerful line of John Deere tractors ever built – the all-new 9030 Series. Choose from 11 models – 4-wheel-drive, scraper special, or tracks.

Big Time Productivity flows from the new John Deere PowerTech Plus™ engine that provides enormous pulling power plus outstanding fuel economy. And with a 10 percent power bulge, the 9630 gives you up to 583 peak horsepower.

On tracks models, the revolutionary new AirCushion™ suspension system isolates the track undercarriage from the tractor's frame giving you a super-smooth ride.

Add a GreenStar™ guidance system and you can make every pass even more productive with lower input costs, minimal overlap, and faster working speeds.

Big Time Versatility – you've got it with multiple drawbar choices and an optional 3-point hitch with up to a 15,300-lb. lift capacity. You can even add up to 6 SCVs. It's extra capacity and versatility for planting, seeding, or tillage.

Big Time Dependability. Since their launch in 1996 John Deere 9000 Series Tractors continue to be the best selling tractors in their class. Why? The answer is simple – dependable pulling power. Visit your John Deere dealer today and check out the Big Time performance of the new 9030 Series Tractors.

*Factory observed with power bulge.

Model	Rated**/Peak Engine hp	Engine Size
9230 4WD	325/357 hp	9.0 L
9330 4WD	375/412 hp	13.5 L
9430 4WD and Scraper Special	425/467 hp	13.5 L
9530 4WD and Scraper Special	475/522 hp	13.5 L
9630 4WD and Scraper Special	530/583 hp	13.5 L
9430T Tracks	425/467 hp	13.5 L
9530T Tracks	475/522 hp	13.5 L
9630T Tracks	530/583 hp	13.5 L

**Factory observed hp levels at 2,100 rpm rated speed. Peak hp figures are factory observed with power bulge.

JOHN DEERE
Nothing Runs Like A Deere®
www.JohnDeere.com/9030

1. The leaping John Deere buck is one of the most appealing and well-known symbols in all of American business. This commemorative sign displays every example since the logo was registered in 1876. Progressive simplification allowed the logos to be more easily stenciled onto Deere products.

1. While John Deere is the producer of a wide variety of products, it remains best known for tractors and other items of agricultural equipment, and is the world's largest manufacturer of such products.

2. Deere's largest tractors come with rubber tires or rubber tracks. There are four-wheel drives with and without articulation, and provisions for as many as three tires on each side of each driving axle. All engines are diesel and range from 61 cubic inches of displacement to 765 cid. Three-, four-, and six-cylinder engines are available. Turbos and intercoolers are not uncommon.

CONTINUED GROWTH IN THE 21ST CENTURY

In 2002, a respected business newspaper, *Crain's Chicago Business*, announced one finding of a nationwide survey: The most trusted Illinois company was John Deere. And in the same year, *Business Ethics* magazine named Deere & Company one of its 100 Best Corporate Citizens. A few years later, in 2007, *Ethisphere* magazine included Deere on its list of the World's 100 Most Ethical Companies. In a period that uncovered awful corporate malfeasance in American banking, energy, and other sectors, Deere's hard-won reputation was golden, indeed.

The firm's reputation for honesty and integrity is particularly noteworthy because of Deere's now-international reach. Major operations are based in Latin America, East Asia, and Australia, and the company maintains its aggressive European manufacturing and sales presence. Today Deere represents not just itself, but the United States, to customers around the globe.

1. Segments of the Agricultural Business side of John Deere include: Greenstar, a satellite-based, computerized piece of equipment that combines navigation function with seed and fertilizer application-rate information, plus many more high-tech solutions to farm-management problems; combines, cotton harvesters, sugar beet harvesters, and hay and forage harvesters; nutrient applicators and sprayers; tillage equipment, planters, and seeders; material-handling equipment, scrapers, cutters, and shredders; gator utility vehicles; diesel engines; home-workshop hand tools for farm and home; frontier equipment; and small tractors and attachments, from landscaping to loader work.

1. Farms grew larger and larger as the millennium approached, and John Deere stayed current with fresh examples of seriously large tractors. The 9300 was one of them. Produced from 1996 to 2002, it tortured the scales at 31,444 lb. The articulated chassis was moved via all-wheel drive and a 765-cubic-inch six-cylinder turbo-diesel that developed 360 horsepower. Maximum fuel capacity was 270 gallons, sufficient for hours of nonstop work. The Synchro gearbox gave 12 forward gears and three reverse, while the PowerSync provided 24 forward and six reverse. A Class 3 hitch meant that the 9300 could pull the heaviest implements without taxing itself.

2. The 9300 Series that bowed in 2007 offered wheeled and tracked versions that produced as much as 583 horsepower.

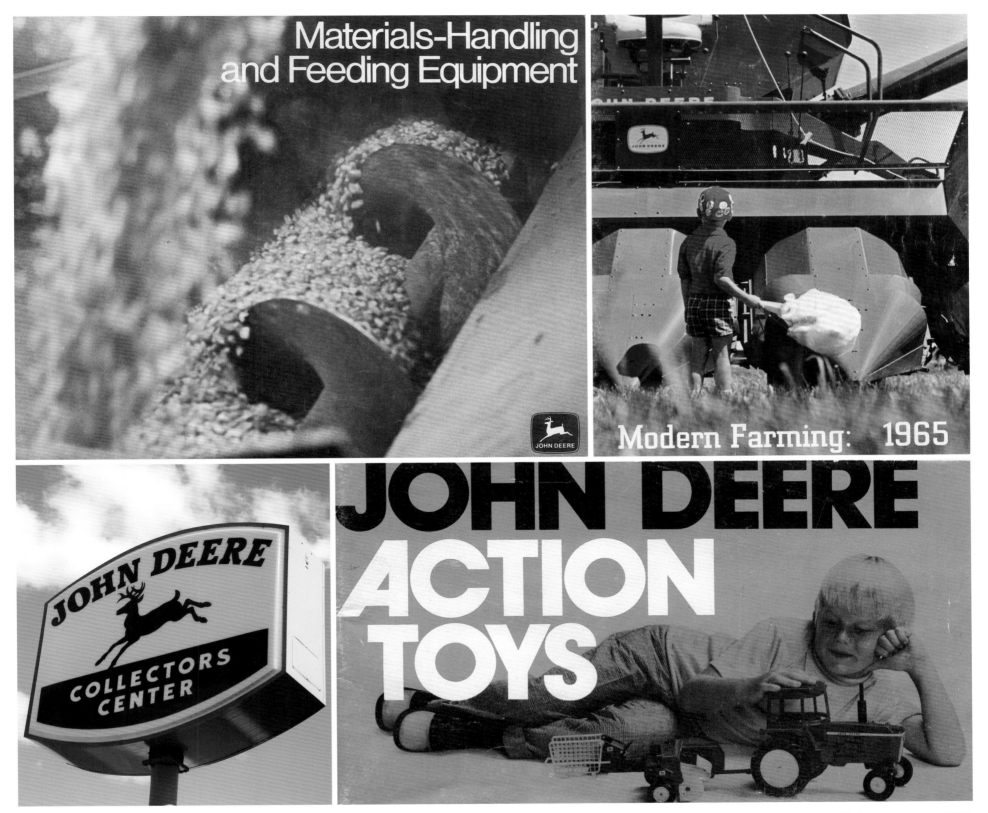

Materials-Handling and Feeding Equipment

Modern Farming: 1965

JOHN DEERE ACTION TOYS

JOHN DEERE COLLECTORS CENTER

INDUSTRIAL EQUIPMENT

Deere's Industrial Equipment Division was established in 1956, along with a separate dealer and marketing organization. After that, a separate engineering department was added. In 1962, a completely new line of products was unveiled: the JDs. As you might guess, all model designations began with the letters "JD." The line now included motor graders, logging devices, loaders, backhoes, excavators, and crawlers. The division now challenged Deere's old friend, Caterpillar, in all quarters.

Caterpillar wasn't Deere's only rival in this segment. To the contrary, Deere simply wanted a piece of a growing pie. International Harvester and Case had complete lines of construction equipment, and there was foreign competition from Komatsu, Kubota, and Mitsubishi.

Log harvesting became an easier job last summer when we introduced our new JD640 Grapple Skidder. That's not just because the grapple eliminates most of the difficult work of attaching cables to trees and sawlogs. We also put operation of the grapple under remote pushbutton control. The operator stays in his protective compartment while he picks up the load. Rounding out our line to six skidders, this one also marks our tenth year of engineering and marketing these modern 4-wheel-drive logging machines.
Your inquiries about John Deere products for forestry and construction and about the company that builds them are welcome. John Deere, Moline, Illinois 61265.

JOHN DEERE on the move

1. By the 1950s Deere tractors had become common in the logging industry. The 440-C was adapted from the 435 for that use, and was seen with wheels or crawler treads. Because of the rugged nature of logging, many of the 440's parts were upgraded for durability. A cast iron radiator grille and extra-sturdy hood helped to protect the engine from damage. For additional security, filler caps were moved under the hood. Instead of the appliqués worn by most John Deere tractors, the "John Deere" name was formed into a steel plate affixed to the sides of the 440-C. A 106-cubic-inch, two-cylinder diesel resided under the reinforced hood and was coupled to a four-speed gearbox. PTO horsepower was 32.91, and enabled the 440-C to confidently move earth and large sections of fallen timber.

2. Startup funds for what morphed into Deere's highly sophisticated "Walking Machine" prototype came from military interests in the Finnish government in the late 1980s, when a Finnish company, Plustech Oy, was tapped to develop a prototype. By 1998, work was based at the Tampere, Finland, headquarters of Timberjack, a Deere subsidiary that controlled Plustech. When Deere acknowledged its connection to the six-legged machine in 2002, the Walker was described as a delicately stepping harvester that could fell timber with minimal trauma to the forest floor (particularly roots and young trees). The articulated, computer-controlled hydraulic legs worked independently of each other, with appropriate placement of each rubber "foot" directed by sensors. The operator controlled the amount of pressure any of the feet brought to bear on the earth, as well as the height of each step. As terrain dictated, overall ground clearance could be adjusted, too. Besides forward and backward movement, the Walker could shift itself diagonally and from side to side. Though remarkable, the Walker ultimately stalled at the prototype stage.

3. The 624J diesel loader has a bucket capacity (heaped) of 3.5 cubic yards. The bucket is 105.9 inches across, and within its full range of motion—full-height extension to ground level—it has a load capacity of 17,290 lb to 29,200 lb. The U.S. military runs a variant called the TRAM.

COMBINES & BALERS

John Deere tractors have long played a crucial role on farms across the country, and joining the tractors were the related hardware that made farming life easier. John Deere's first combine appeared in 1927 and was named the Number 2. Further innovations created the Number 1, which appeared in 1928. Both of these were surpassed by newer and improved models. The first self-propelled models arrived in 1947. Continual revisions brought new innovations every year, and sales grew as the improving capabilities of John Deere's combines made quicker work of harvesting chores. In 2010, John Deere produced their 500,000 self-propelled combine, setting standards for the industry.

In 1900, John Deere was buying balers from Dain Manufacturing Company in Ottumwa, Iowa. In 1910, Deere bought the Dain company outright as sales of the popular device continued to grow. Several decades later, John Deere had a record year in the baler business with its 14-T model. Improvements in the technology reduced manual labor by 50 percent and added more efficiency to the farming efforts. Today's 900 Series of John Deere balers can deliver levels of performance that were scarcely imaginable 100 years ago.

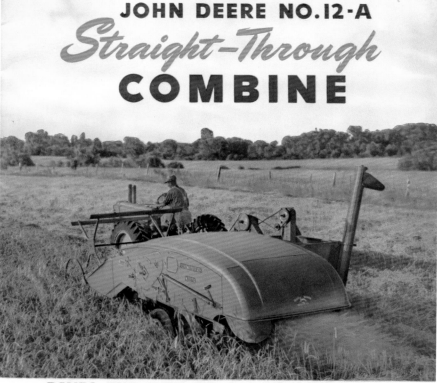

You Get

MORE FOR YOUR MONEY

In a JOHN DEERE *Straight Through* COMBINE

THE extra value that's built into John Deere *Straight-Through* Combines reflects itself in shorter, easier and more profitable harvests.

In John Deere Straight-Through Combines, grain and straw are handled in a straight line from the cutter bar on through the machine. There are no turns, no corners to cause clogging or piling. Big-capacity, rasp-bar cylinder . . . full-width separation, and extra-large cleaning units insure faster, cleaner threshing in all crops.

Heavy-duty, canvas-type platform which cuts from 1½ to 40 inches from the ground . . . slip-clutch, ground-driven reel . . . low-down, auger un-loading grain tank . . . V-belt drives . . . simple sliding hitch for narrowing down transporting width . . . high-grade bearings throughout . . . a minimum of grease fittings, located where they are easy to reach . . . an even, uniform spread of straw behind the combine . . . and safety slip clutches wherever needed are other advantages that pay big dividends in John Deere Combine ownership.

Take the time now to write for free folder covering the No. 11-A five-foot and the No. 12-A six-foot combines. Then arrange to see these great cost-reducing combines at your John Deere dealer's.

STANDING GRAIN AND SORGHUM COMBINES

The New JOHN DEERE 25 COMBINE

WITH 6- OR 7-FOOT CUT

It's a Specialist in All Crops

NEW FEATURES INCLUDE: Quick-change cylinder speed control...
More aggressive open-bar grate with snap-in inserts...
All-steel straw rack...Improved cleaning shoe...
25-bushel grain tank...Stronger frame and hitch...
Better flotation for soft or muddy fields...
and many other valuable improvements

RECREATION

No one lives by farming alone. To aid leisure-time pursuits, John Deere makes a variety of recreational and light-duty utility products for the consumer. In 1963 a new Deere Consumer Product Division was established at the old Van Brunt factory in Horicon, Wisconsin. Its first product was the immensely popular Model 110 lawn tractor. The line developed to include, at least for a time, bicycles, snowmobiles, all-terrain vehicles, and chain saws, but now concentrates on small tractors, mowers, snow blowers, and other items for the homeowner.

Today, the ZTrak Zero-Turn riding mowers offer the convenience of a tight turning radius to make lawn mowing fun. For bigger yards, the company makes a huge variety of lawn tractors that start with 17.5 horsepower and can carry mower decks of up to 60 inches in width. A broad line of Gator utility vehicles includes everything from the basic utility model up to the high-performance RSX860i, which is powered by a 62-horsepower v-twin engine and can hit a top speed of 60 mph.

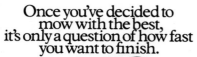